Microsoft®
Word 2010
Beginner's Guidebook

by Allen L. Wyatt

tips.NET

Microsoft® Word 2010 Beginner's Guidebook

Published by:

Sharon Parq Associates, Inc.
PO Box 794
Orem, UT 84059

For information on purchasing, distributing, or reselling books published by Sharon Parq Associates, Inc., please visit our website (www.SharonParq.com) or call 801-607-2035. Our books are also available through select online resellers such as Amazon.

ISBN: 978-1-61359-069-0 (printed book)
ISBN: 978-1-61359-070-6 (e-book on CD-ROM)
ISBN: 978-1-61359-071-3 (e-book)

Produced and published in the United States of America

Revision history:

2 November 2011: First published

Content Overview

About the Author

Allen Wyatt, an internationally recognized expert in small computer systems, has been working in the computer and publishing industries for more than three decades. He has written more than 50 books explaining many different facets of working with computers, as well as numerous magazine articles. His books have covered topics ranging from working with programming languages to using application software to using operating systems. Through the written word, Allen has helped millions of readers learn how to better use computers.

Besides writing books, Allen has helped educate thousands of individuals through seminars and lectures about computers. He has presented complex topics to audiences throughout the United States as well as throughout Mexico and Costa Rica. His books, which often form the basis of his presentations, have been translated into many languages, including Chinese, Dutch, French, German, Greek, Italian, Japanese, Korean, Polish, Russian, and Spanish.

Allen is the president of Sharon Parq Associates, Inc., a computer and publishing services company located in Orem, Utah. Besides writing books and technical materials, he publishes a series of online newsletters and oversees the development of the Tips.Net family of websites (www.tips.net).

Allen Wyatt
Sharon Parq Associates, Inc.
PO Box 794
Orem, UT 84059

allen@sharonparq.com

Detailed Table of Contents

Introduction

Everyone knows that Microsoft Word is the best-selling, most-used word processing program in the world. (You did know this, didn't you? If not, just go along with me on this—I don't want to have to pull out all sorts of sales data to dazzle you.) Most new PCs being sold these days come with some version of Microsoft Word already installed on them, and a level of fanfare normally reserved for kings and potentates accompanies each new release of the software. Word is reigns supreme among word processing software.

Just because a piece of software is dominant does not mean that it is easy to use, however. Word 2010, the latest version of the software, has more features, bells, and whistles than you can shake a stick at. Many people, when faced with the challenge of learning how to use Word, are overwhelmed by the sheer volume of program capabilities.

Word 2010 presents an additional challenge, however. Long-time users of Word may just now be upgrading to Word 2010, having skipped the previous version of the software (Word 2007). Why is this important, besides the obvious fact that it affects Microsoft's bottom line? Because Word 2007 introduced the totally new user interface that is also used by Word 2010. This means that long-time Word users almost start over, from scratch, in learning how to use the program.

Thus, the need for ***Microsoft Word 2010 Beginner's Guidebook.*** This book provides the essential information you need in order to start using Word 2010 effectively—right away. The information presented in this book is presented in bite-size sections that had their genesis in tips presented in the weekly ***WordTips*** newsletter. (You learn more about ***WordTips*** shortly.) This approach—using bite-size sections—is the ideal way of learning how to use Word. Rather than overwhelm you with everything you can do, the focus is on how to accomplish common tasks quickly and easily.

Who Is This Book For?

If you are new to Word, ***Microsoft Word 2010 Beginner's Guidebook*** is for you. If you don't know a style from a watermark or a template from a list, you will definitely benefit from reading this book. If, on the other hand, you are very comfortable using Word 2010 to create reports, write letters, format tables, and crank out labels, then perhaps this isn't the book for you. A quick skim through the next section will help you to make your final decision, however.

This book is also handy if you're training others in how to use Word. Hundreds of computer trainers and help-desk personnel have benefited from using *WordTips,* and this book is designed to make their lives much simpler when working with Word newcomers.

There are a couple of assumptions that I make about you. First, I assume that you already have Word 2010 installed on your computer system. Second, I assume that you know how to do elemental tasks, such as start Word and exit the program. Finally, I assume that you are not afraid to experiment and learn exactly how to do what you want to do in Word.

What Is in This Book?

This book contains quite a bit of information essential to learning how to use Word 2010. Don't make the mistake of thinking that it contains information on every possible way you can use Word; it does not. In fact, several handy features in Word aren't covered at all. That is okay, however, because knowing how to use those features is best left to a later book and learned after you master the essentials of Word.

You should view ***Microsoft Word 2010 Beginner's Guidebook*** as an introduction to the wonderful world of Word. Here, you learn "the lay of the land" and how to get around and how to start using Word productively. In this book's ten chapters, you learn about the concepts I describe here.

Chapter 1, "Getting Acquainted with Word 2010," introduces the basics of how to survive with Word. Here, you learn about such elements as the status bar, views, document windows, shortcut keys, and getting help when you need it.

Chapter 2, "Configuring Word 2010," discusses how to modify how Word behaves. You learn how to change not only Word's appearance, but also what you see in your document, how editing works, and how formatting is applied.

In Chapter 3, "Creating and Editing Documents," you learn how to "crunch text" with Word. (This is, after all, the heart of word processing, right?) You learn the basics of navigating, selecting text, editing text, copying and moving text, working with headers and footers, and using page numbers.

Chapter 4, "Searching and Replacing Information," introduces the powerful find-and-replace feature of Word. Here you learn how to search for not only text, but also other types of special characters and formatting.

Formatting is discussed in Chapter 5, "Formatting Your Document." You learn how to apply formatting in each of the different levels of your documents: pages and sections, paragraphs, and characters. You also learn how to work with tab stops, columns, line numbers, and lists.

One of the most powerful features of Word is expressed in a single word: styles. Chapter 6, "Introducing Styles and Templates," covers the basics of how to put styles to work.

In Chapter 7, "Using Tables," you learn how to master Word's table editor. Learn how to create, change, reconfigure, format, and delete tables. You also find out about some handy shortcuts that can speed up your use of table data.

Chapter 8, "Working with Fields," discusses how you can use fields to insert specialized and dynamic information in your documents. Examples are provided using the most common types of fields you may encounter.

Chapter 9, "Printing Documents," covers all the basics of putting your documents on paper. You learn how to control the printer, set up a print job, and prepare documents for special needs.

In Chapter 10, "Managing Documents and Files," you learn a few tips that may save you tons of time when you're using Word to load, save, and manage files.

A Note about WordTips

As I mentioned earlier in this introduction, much (but not all) of the information in this book comes from a publication named ***WordTips***. This electronic newsletter has been published weekly since early 1997. Subscriptions to ***WordTips*** are free; all you need is an e-mail address at which to receive your issues. If you want to sign up and receive ***WordTips*** for yourself, you can do so at the following Web address:

```
http://wordribbon.tips.net
```

I also publish a sister publication, ***ExcelTips***. If you use Excel, you will find the information in ***ExcelTips*** just as valuable as the information in ***WordTips***. To receive your own issues, visit this Web site:

```
http://excelribbon.tips.net
```

1

Getting Acquainted with Word 2010

Welcome to Word 2010. Hopefully, you have had an opportunity to sit down and try to use Word a couple of times. As you worked with the program, you may have had some awkward moments as you tried to figure out what to do next. If you have not used Word before, you may be hesitant to jump right in and start using the program.

Never fear; Word isn't that difficult to use if you simply learn a few basic lessons about how to use it. The tips in this chapter are geared toward the most elemental portions of the Word environment. Here, you learn about different parts of the Word program window, how to use views, how to view different parts of your document, what shortcut keys Word has to offer, and how to get help in Word. You even learn some tips that are very helpful if you are migrating to Word from a different type of word processor.

Understanding the Basics of the Word Environment

When you start Word 2010, the majority of the program window is taken up with your document—where you start typing your next masterpiece. There are other elements that are visible in the program window, however, and understanding these elements is crucial to effectively using Word.

The following sections describe the various parts of the program window and how you can use them effectively. You specifically discover the ribbon, the Quick Access Toolbar, and the status bar.

Introducing the Ribbon

Word 2010 uses what is known as a ***ribbon-based interface***. This interface is different than what you may be used to in other programs or even in past versions of Word. A ribbon-based interface doesn't rely on traditional menus, but instead makes available tabs that present collections of tools that can be used to accomplish related tasks.

In some ways, the ribbon is similar to the tabs in a dialog box. Each tab makes tools available, and all you need to do to display those tools is to click on the tab you desire.

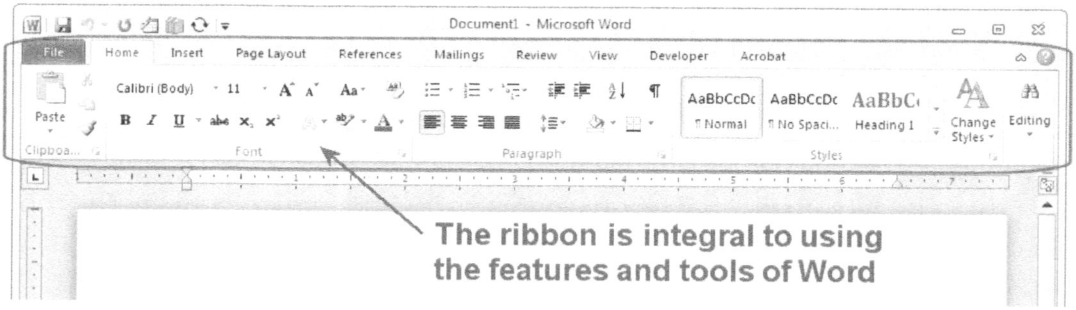

The tabs displayed on the ribbon are contextual in nature. This means that the tabs available depend on what you are doing in your document at the current time. Many of the tabs (such as File, Home, Insert, etc.) are nearly always visible. Other tabs are only visible when you are working in objects where it makes sense to display the tab. For instance, you may only see the Design tab when you are working in a table.

A Special Case: The File Tab of the Ribbon

As you work with different ribbon tabs, your document is normally still visible. Granted, clicking a particular tool may display a dialog box, a palette, or a drop-down list, but you can still see your document patiently waiting in the background of your program window.

There is one ribbon tab that doesn't behave in the normal way; it completely hides your document while it is displayed. This is the File tab of the ribbon. You'll note that this tab is displayed in blue, unlike all the other ribbon tabs. When you click on it, your document disappears and the entire screen is taken up with the options made available through this tab.

Also unlike other tabs, you can return to your document by simply pressing the **Esc** key. Word once again displays your document along with whatever ribbon tab was visible before you clicked the File tab.

You should take some time to explore the various options made available through the File tab of the ribbon. It is the gateway to important tasks, such as creating and saving new documents, printing your current document, and configuring how Word works. In fact, as you work through the chapters of this book you'll discover many times when you need to work with the File tab.

Displaying the Full Ribbon

The ribbon can either be fully displayed or it can be minimized when not in use. You may find it very helpful to minimize the ribbon so that you can use more of the screen area to work with your document. You can minimize the ribbon by using any of these methods:

- Press **Ctrl+F1**.
- Right-click any ribbon tab and choose the Minimize the Ribbon option.
- Click the small upward-pointing arrow at the right end of the ribbon.

When you minimize the ribbon, only the names of the ribbon tabs are visible. You can display the tools on a tab by clicking once on the name of the tab. When you are done using the tools, they safely minimize so that you can again work in your document.

You can switch the ribbon back to a normal, full display by using any of the same methods you used to minimize it. In addition, you can show the expanded ribbon by double-clicking on the name of any ribbon tab.

The Quick Access Toolbar

Take a look at the upper-left corner of the program window and you'll notice a handful of small icons. These are located just above the File tab of the ribbon and are collectively known as the ***Quick Access Toolbar***.

The purpose of the Quick Access Toolbar is to display the tools you'll use most frequently when working with your documents. By default the Quick Access Toolbar displays tools that allow you to save your document (the icon looks like a small diskette) and tools to undo and redo changes you make to your document. There may be other icons visible, as well, depending on how your system is configured.

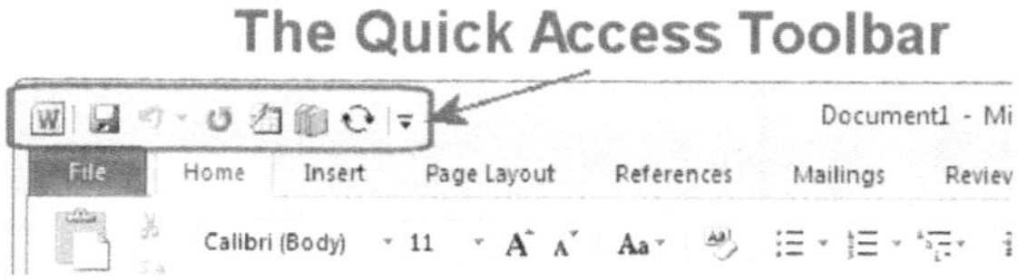

If you want to control what is displayed on the Quick Access Toolbar, click the downward-pointing arrow at its very right side. Word displays a list of common commands, along with indicators of which commands are displayed on the Quick Access Toolbar. Pick a command and it appears; pick it again and it disappears—it's that simple!

The Status Bar

At the very bottom of the program window is the area known as the status bar. This area runs the entire width of the program window and is typically used to display status information about you document. (Thus the name "status bar." Bet you didn't see that one coming!)

You can control what is displayed on the status bar by right-clicking someplace within the status bar that doesn't contain any information. (Someplace right in the middle of the status bar is a good choice.) Word displays a list of information

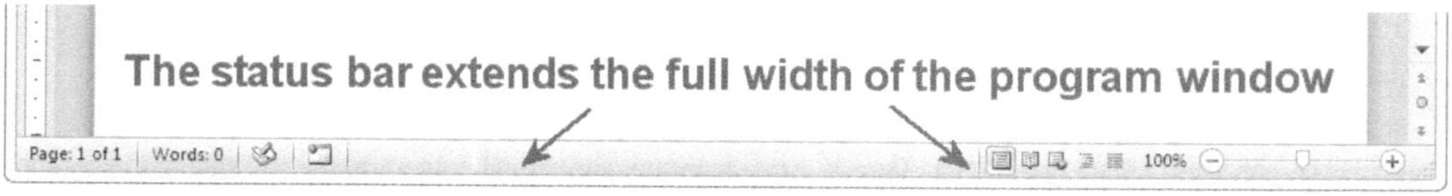

you can display, along with an indicator of whether that information is currently being displayed. If a check mark appears beside the information, then it is currently visible. You can turn the display of an item on or off by clicking where the check mark is located.

Displaying Your Document

Once you understand the basics of the Word environment, you have a firm foundation on which to build your set of Word skills. The following sections describe, in detail, some basic concepts you need to understand relative to displaying your document. Here, you learn how to work with views, document panes, windows, and nonprinting characters. You also get a crash course in how to use the function keys that Word makes available for your use.

Understanding Views

Word provides different ways you can view your documents, depending on your particular needs. The major views available in Word are Print Layout, Full Screen Reading, Web Layout, Outline, and Draft. You can choose which view you are using by clicking on the view controls at the right side of the horizontal scroll bar or by selecting a view from the View tab of the ribbon.

Print Layout view is the view most closely related to what your document will look like when you actually print it. This viewing mode, which is the one you will probably use most of the time, allows you to see your headers and footers in place, what your margins look like, how your text boxes appear in relation to text, and what your graphics look like in your document. This is the viewing mode you should use if you want to always see what your document will look like.

Full Screen Reading view allows you to do exactly what its name suggests—read your document using the full screen of your system. The view gets rid of the ribbons and uses the maximum screen space available to display your document. Typically the document will be displayed in two facing pages, but the number of pages displayed can be affected by the size of the monitor you are using. There is no editing allowed in this view; it is for reading and

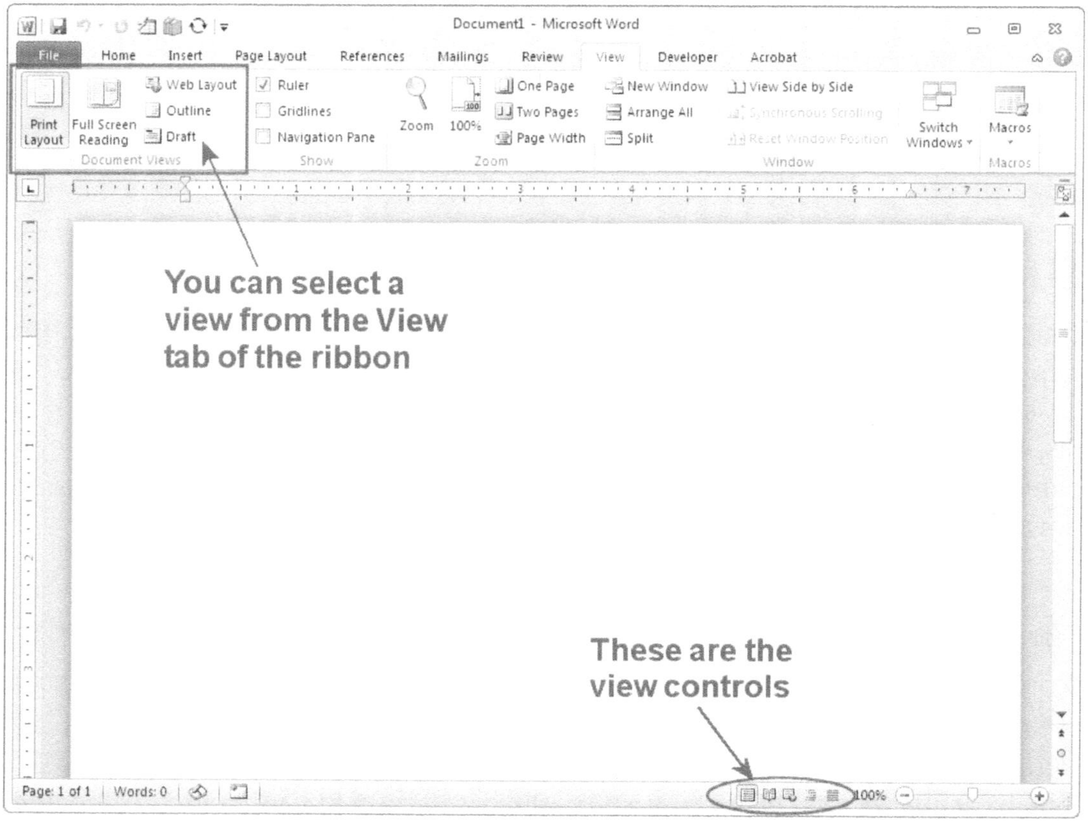

reviewing only. You can exit this view by simply pressing the **Esc** key. (In many ways, Full Screen Reading is very similar to Print Preview in older versions of Word.)

Web Layout view is designed to allow you to easily see how your documents will look if used in an online environment. There is not much more to say about this viewing mode; it is provided for those who intend on publishing their Word documents online.

Outline view is used when you want to work with large portions of your document at the same time. It allows you to collapse your document and view only the major headings. The text under each heading can be hidden so it does not obscure your view of document organization. When you select Outline view, an additional Outlining tab appears on the ribbon. This tab allows you to control what is displayed in this view.

Draft view can be considered a "pared down" version of the Print Layout view. It allows you to generally see how your text will appear on paper. This means you can see what each line will look like, how the text appears, and where the lines will break. You can also see where each page will break. This view is helpful if you are using an older, slower computer that can't display the Print Layout view particularly quickly. (Print Layout view requires more computing overhead to display information.)

If you open multiple documents or use multiple panes to view the same document, switching views in one of the windows or panes will not affect the others. Word controls the panes independently; thus, you can use one document pane to see what your document looks like in one view and another to work with the document in an entirely different way.

Working with Document Panes

Word allows you to divide a document window into two ***panes*** that allow you to view two different parts of the same document. To divide a document window into panes, you use the divider bar located in the upper-right corner of the document window. The divider bar is located at the top of the vertical scroll bar on the right side of your document; it looks like a solid horizontal bar. When you position the mouse pointer over the divider bar, it changes to a different type of pointer—it has two heads pointing up and down. Click on the divider bar and drag it to where you want the document window divided. If you want to divide the document window in half, you can simply double-click on the divider bar.

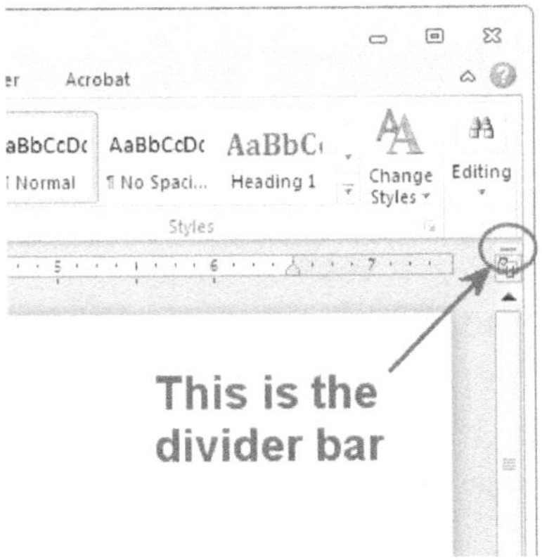

You can also create different panes by displaying the View tab of the ribbon and then clicking the Split tool. Word places a horizontal split bar on the screen that you can easily position by moving the mouse. When you click on the mouse button, the window splits into panes exactly where you clicked. If you want to cancel the split, just press **Esc** before you click the mouse button.

After dividing a document window into panes, you can adjust the size of the panes at any time. You do this by clicking and dragging the divider bar to the new vertical location where you want the document window divided. When you release the mouse button, the divider bar remains at that point. Because each pane is an independent view of your document, you can use different views in each pane—for instance, Draft view in the top pane and Print Layout in the bottom.

If you want to get rid of the panes, you can return to a single window in two ways. Try either of these methods:

- Click the Remove Split tool on the View tab of the ribbon.
- Double-click on the split bar.

When you remove the panes, the resulting single window maintains the attributes of the pane in which the insertion point was located. Suppose that the pane above the split bar shows your document in Draft view and the one below the split bar shows your document in Print Layout view. If the insertion point is below the divider bar when you remove the panes, the resulting single window will be in Print Layout view. However, if the insertion point had been above the divider bar, the resulting window would have been in Draft view.

Creating New Windows

If you want to work on two different parts of the same document at the same time, you can do so in Word in a couple of different ways. One way is to open a second window. You do this by displaying the View tab of the ribbon and clicking the New Window tool. Word opens a new window and you can use each window to display and edit different parts of the same document.

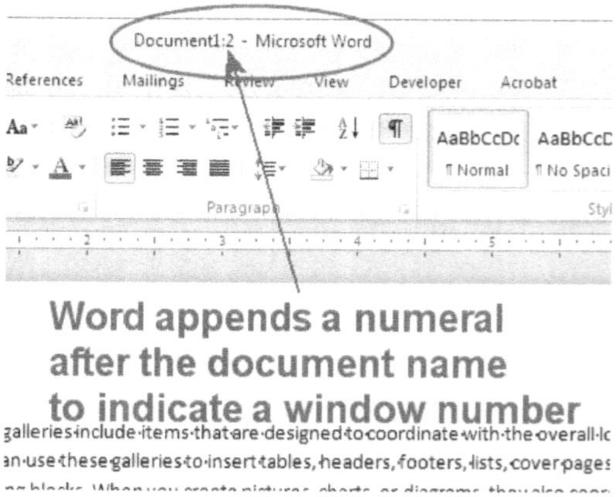

Notice that each new window you create has not only the document name on the title bar, but also a number that indicates the window number. Thus, you could have MyDocument:1 and MyDocument:2.

Each window created in this way just provides a different way to look at exactly the same document. Any change you make in one window is automatically and immediately made in the other window as well.

Arranging Document Windows

If you have more than one document open at a time, Word allows you to view all the documents at the same time, and to arrange each document window as you desire. The easiest method of arranging document windows is as follows:

1. Display the View tab of the ribbon.
2. Click the Arrange All tool, in the Window group. This divides the screen space evenly between all your windows. The main benefit is that all your document windows get at least a piece of the screen.
3. Click on the window whose size you wish to adjust. It becomes active.
4. Move the mouse cursor near the border of the active window. When the cursor turns into a set of arrows, press the mouse button and drag the window edge to the desired size. Release the mouse button.
5. Repeat steps 3 and 4 for each window you want to adjust.

Understanding Nonprinting Characters

Word keeps track of many characters that don't normally show up on the screen or on a printed copy of your document. You use many of these characters every day, but probably don't think of them as characters (as such). The nonprinting characters that Word uses include the following:

- Hidden text
- Paragraph marks and newline characters
- Optional hyphens
- Column, page, and section breaks
- Spaces
- Tabs

Each of these can be individually displayed in Word by displaying the Word Options dialog box—display the File tab of the ribbon and then click Options at the left side of the screen. Finally, click Display at the left side of the Word Options dialog box.

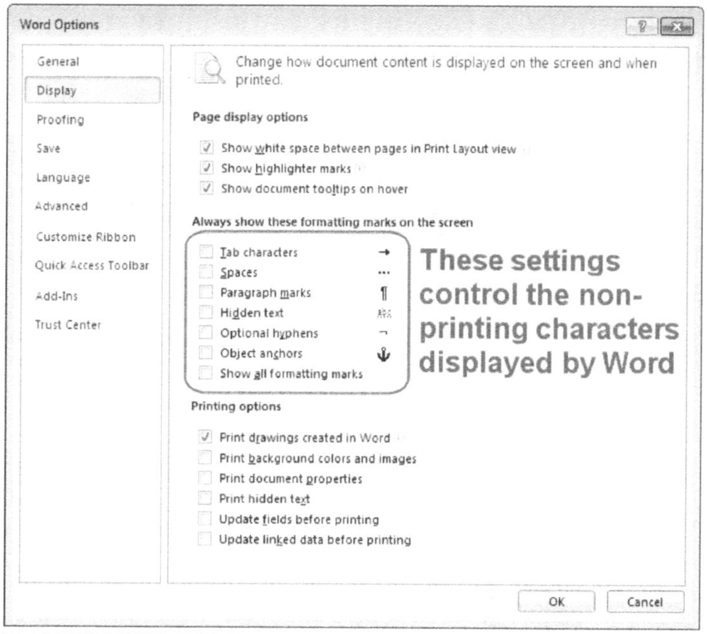

Sometimes when you are working in Word, it is beneficial to make sure that nonprinting

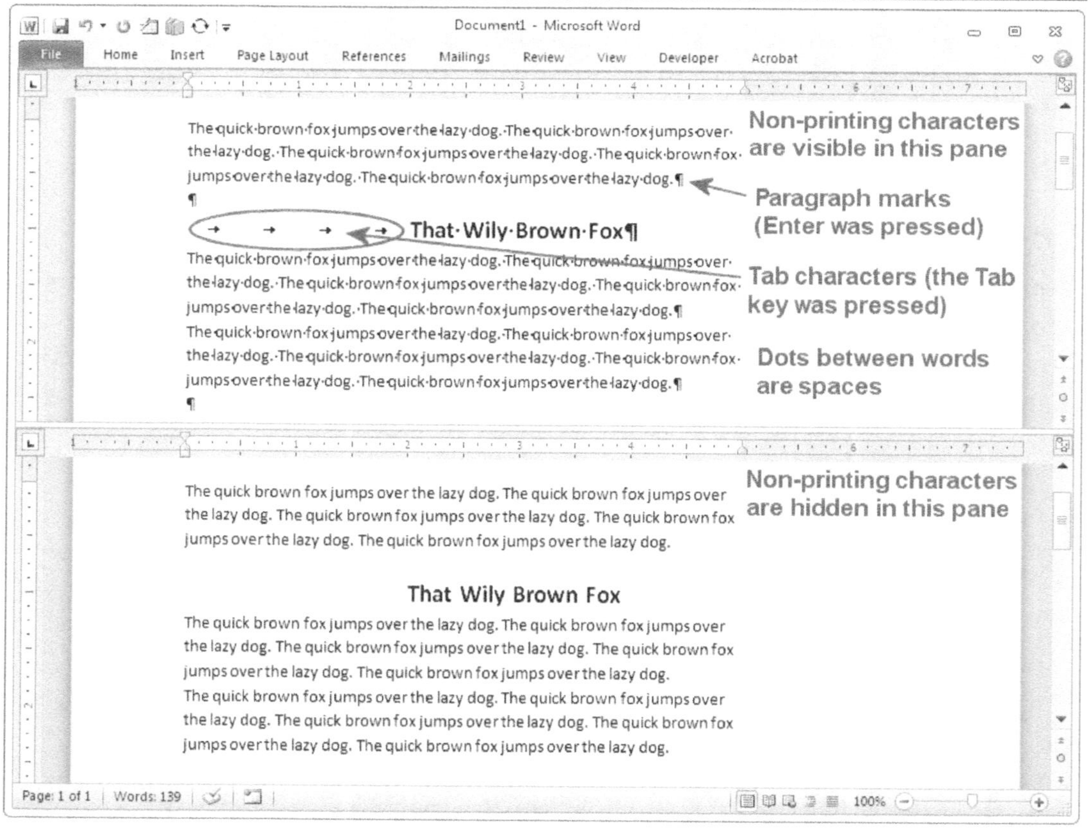

characters are displayed. For instance, if you cannot seem to get something aligned properly or formatted just right, nonprinting characters may be the culprits. Display them, and then try your formatting.

Word Function-Key Assignments

Many *WordTips* readers have asked for a definitive list of function keys and their uses. The table on the next page provides such a list.

Getting Help in Word

It makes sense that at some time you need help while you are using Word. Your refuge should be the help system built into Word. You can access the help system by using any of these methods:

- Press **F1**
- Click the question-mark-within-a-circle icon at the upper-right corner of the program window
- Display the File tab of the ribbon and click Help at the left side of the screen

The level and type of help displayed through each of these methods is different. The following sections provide further information on how you can get help from Word exactly when you need it.

Getting Context-Sensitive Help

Word includes a complete help system that you can use to answer most questions on how the program operates. Because this system is ***context-sensitive,*** the help you receive varies depending on the context in which the help was requested. Word provides a fast way to ask for context-sensitive help. To do this, press **F1**. Word displays the Word Help dialog box where you can further refine what you are looking for.

	Plain	Shift	Ctrl	Alt	Shift+Ctrl	Shift+Alt	Ctrl+Alt
F1	Invoke help system	Reveal formatting		Next field		Previous field	Display Microsoft System Information
F2	Move selection	Copy selection	Print preview			Save	Open
F3		Change case of selection	Cut selection to the Spike	Create a Building Block entry	Paste information from the Spike		
F4	Repeat last action	Repeat last Find or Go To operation	Close window	Quit		Quit	
F5	Go To	Return to previous edit location		Restore program window size	Edit bookmark		
F6	Next pane	Previous pane	Next window	Switch between dialog box and document	Previous window	Previous window	
F7	Spell check	Thesaurus		Select next spelling error and display Context menu	Update links	Display Research task pane	
F8	Extend selection	Collapse selection		Display Macros dialog box	Start COL selection mode		
F9	Update fields	Toggle display of field code or results	Insert field code braces	Toggle display of all field codes or results	Convert field to regular text	Run GOTOBUTTON or MACROBUTTON from a field	
F10	Show Key Tips	Display Context menu	Maximize document window	Maximize program window			
F11	Next field	Previous field	Lock a field	Display VBA editor	Unlock a field		
F12	Save As	Save	Open		Print		

Getting Help in Dialog Boxes

As with most Windows programs, Word uses dialog boxes to present information and allow you to make changes to various settings. Sometimes the plethora of options in a dialog box can be rather confusing. If you need help on a particular option within a dialog box, you can use a built-in feature of Windows. Examine the dialog box; in the upper-right corner you see some controls. The most common control is the Close button. Many dialog boxes may also have a button with a question mark on it. If so, you can click on the question mark and Word then displays the Word Help dialog box with further information about the options available to you.

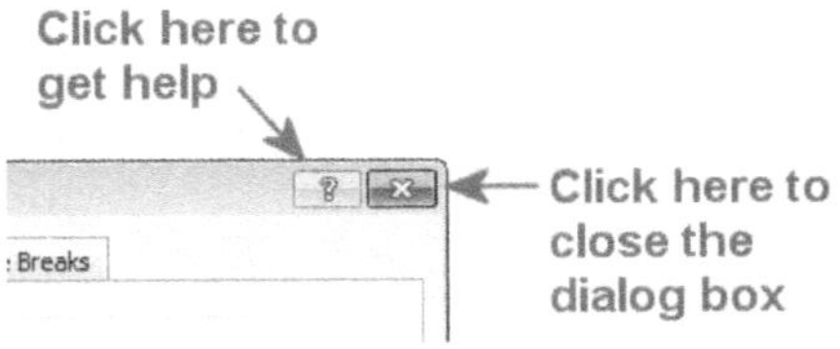

Turning Off Online Help

The default settings for Word always use the online capabilities when pulling up helpful information. You can, if you prefer, instruct Word to ignore any online resources. To change the settings, follow these steps:

1. Press **`F1`**. Word displays the Word Help dialog box.
2. Click the Connected to Office.com wording at the bottom-right of the status bar.
3. Click Show Content Only From this Computer. The dialog box's status bar changes to Offline.
4. Use the Help system, as desired.

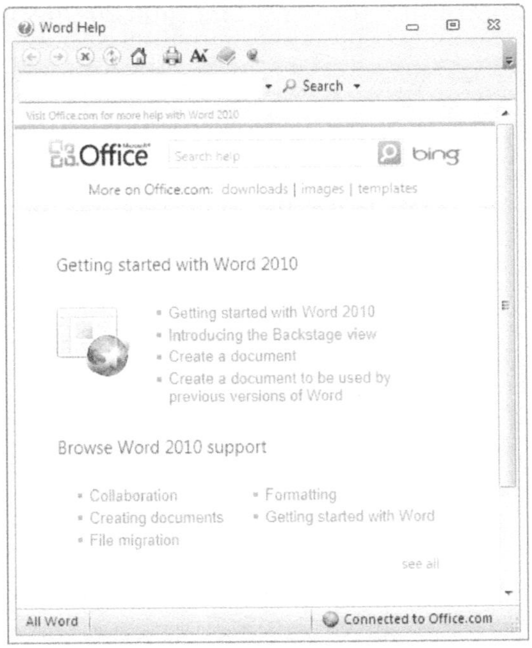

Migrating to Word

Word is probably not the first word processor you have used; Microsoft wants it to be the last, however. (Sorry; the PR folks in Redmond, Washington, made me say that—*not!*) If you are coming to Word from another word processor, you should pay particular attention to a couple of things. In the following sections, you learn how Word names its versions, why it doesn't have reveal codes (as in WordPerfect), how to enter values into a dialog box, and several other nuggets that are quite important.

Come to think of it, these items should be of interest to everyone, not just to those migrating to Word. Take your time and read through these sections; the information they contain comes in very handy in later sections of this book.

Understanding the Word Name Game

Relative newcomers to the wonderful world of Word can easily become confused when discussing Word versions. To clarify any misunderstanding, the following information is offered. (Word history mavens are welcome to contribute any additional information concerning the version history.)

- Word 2 was available on the market in the early 1990s. It was used and abused by millions. I still have several books on my bookshelves that relate to this version of Word. (OK, I admit it—I'm rather slow in clearing out my bookshelves.)
- Word 6 was released in 1993. There were actually four different versions of Word 6 that you may hear reference to: Word 6.0, 6.0a, 6.0c, and Word NT, version 6.0. The version for Windows NT was actually released in 1994.
- Word 95 was released in 1995. This is often referred to as Word 7. (Word 95 and Word 7 are one and the same.)
- Word 97 was released in November 1996. This version is also known as Word 8.
- Word 2000 was released in June 1999. This version is sometimes referred to as Word 9.
- Word 2002 was released in August 2001. This version is sometimes

referred to as Word 10. (Sometimes the version is incorrectly referred to as Word XP. There is no Word XP, but Word 2002 is included in the Office XP suite. Go figure.)

- Word 2003 was released in October 2003. This version is sometimes referred to as Word 11.
- Word 2007 was released in January 2007. This version (can you guess?) is sometimes referred to as Word 12.
- Word 2010 was released to the public in June 2010. Microsoft skipped the Word 13 nomenclature (it might be unlucky) and called this one Word 14.

There were numerous differences between each of these versions, and some versions had multiple service packs released for them, which meant there were even more differences. There are also versions of Word released for other platforms, such as the Mac, and they go by even different names.

Among the versions noted above, the biggest differences from one version to another were with the introduction of Word 97 and Word 2007. Word 97 changed many interface features, the file formats used by Word, and the programming language used for macros. Word 2007 introduced an even more radical change in the user interface along with another new file format, but no drastic changes in the programming language for macros.

On a historical note, you may find it curious that there was a jump from Word 2 to Word 6. The reason was that although pre-Word 6 versions of Word were on the market, versions of Word for DOS were also on the market. Word 5 (for DOS) was the last major version of the product for the DOS world. To show some supposed continuity in the family, Redmond adopted the DOS version numbering for its next major Word product, resulting in Word 6. Some people also correctly point out that the major competitor to Word 2 was WordPerfect 5.1. By bringing out Word 6, Microsoft was perceived to have a "newer" word processor than WordPerfect's.

Working with Reveal Codes in Word

Many people write to me and express frustration because they miss the reveal codes feature they were used to in WordPerfect. They do not understand why

Word has no comparable function. The reason is that there is a core difference between how documents are formatted in Word and WordPerfect, and each takes a totally different approach to text in general. I won't try to convince anyone that one approach is better than the other; both have their pros and cons. Suffice it to say that they are just different from each other.

WordPerfect essentially treats text and formatting codes the same. When you format some of your text, a start code is inserted at the beginning of the selection and an end code is inserted at the end. These codes are actually inserted into the text and the code markers are visible when you display the reveal codes window. You can select and delete these codes, just as you would do with any other character. In fact, you can insert a start code in your text and if you don't insert an end code (or if you delete the end code without deleting the start code), the rest of the document is formatted according to the lone start code.

You can't do that in Word because Word treats text and formatting completely independent of each other. The file format used in Word stores formatting information in a different portion of the file than the text to which that formatting is applied. Thus, the formatting is not within the "text stream," as it is in WordPerfect. Instead, Word tracks the paragraph and character formatting applied to every character in the document, without the need to worry about start codes or end codes.

In Word, basically "what you see is what you get." If something is formatted as bold text, it appears as bold text. It is essentially impossible to have stray or unused formatting codes in your text, as you can in WordPerfect. The reason is that Word has no such codes.

The closest you can get to the reveal codes feature is to display the complete formatting information that Word is applying to a character or paragraph. You do this by pressing **SHIFT+F1**. Word displays, on the right side of the document window, the Reveal Formatting task pane.

The Reveal Formatting task pane contains detailed information about the formatting of whatever you have selected in the document.

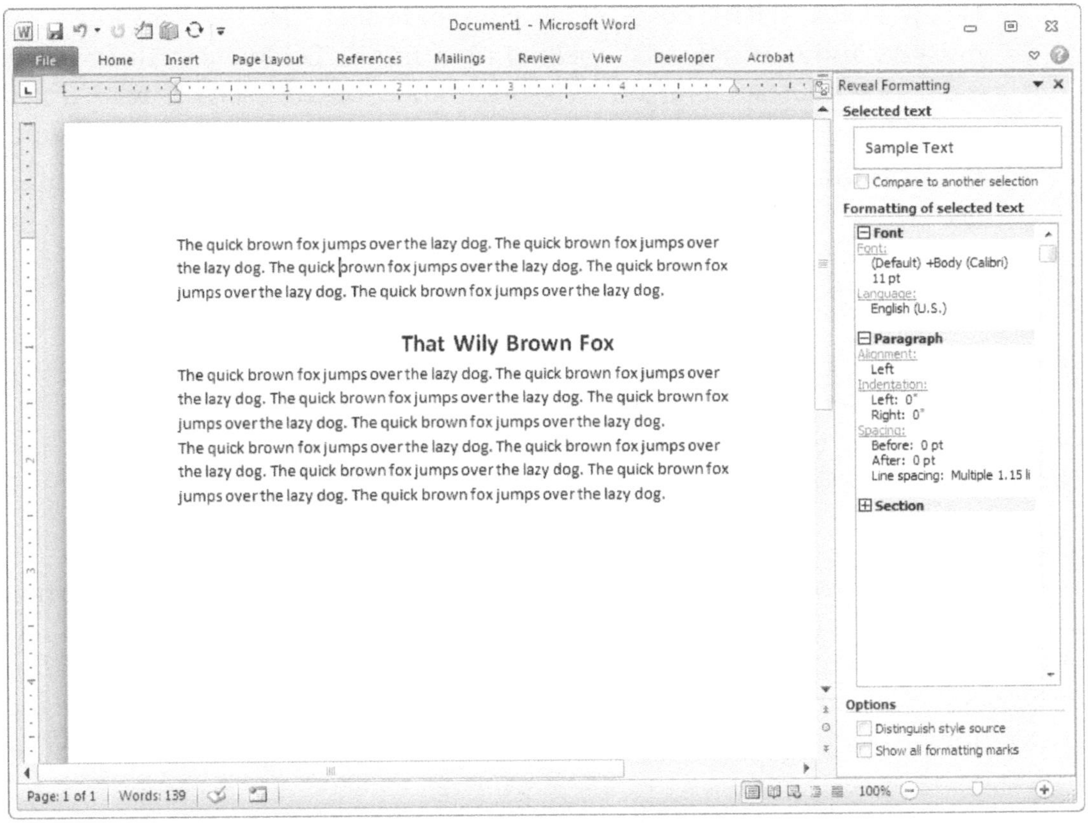

Understanding the Status Bar Icons

As you learned earlier in this chapter, the status bar is the area at the bottom of the program window that allows you to see the status of your document or Word itself. Some items on the status bar are intuitive; others may need a bit of explaining. In this section, you look at the icons on the status bar and what they signify.

The most common icon that appears on the Word status bar is a small, open book. This book appears when automatic spell checking or grammar checking is turned on. The icon can have three different appearances:

- **Open book with pencil.** This icon signifies that Word is busy checking the spelling or grammar used in your document. For instance, start typing something—anything. As you type, the pencil is visible; shortly after you stop, the pencil disappears and is replaced with a different version of the icon.

- **Open book with red X.** This icon indicates that the checking has finished and that potential spelling or grammar errors were discovered. Each potential spelling error is underlined on the screen with a red squiggly line, and grammar errors are noted with a green squiggly line.
- **Open book with check mark.** This icon shows that Word has found no errors in your document. The easiest way to see this icon is to create a blank document and type a single word (make sure you spell it correctly). The icon should show the check mark.

Note that if you have a red X on your book icon, you can locate the next error in your document (along with suggested corrections) by double-clicking on the book icon.

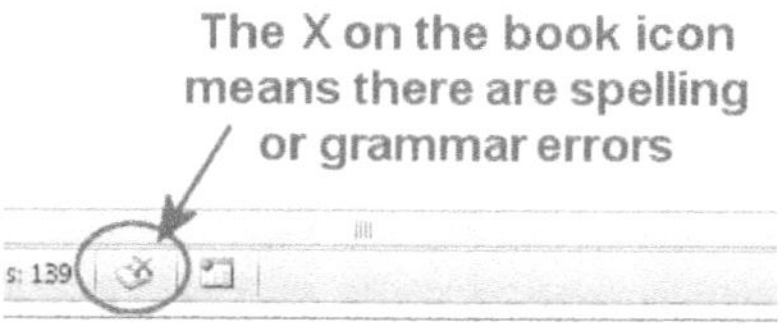

The only other notable icon on the status bar is the one that looks like a printer printing. It is visible if you have turned on background printing and have instructed Word to print a document. As your document is printed to the background spooler, the number on the icon indicates the page being printed. If you double-click on this icon, you have the opportunity to cancel the print job.

Understanding Hard and Soft Returns

When you use a typewriter, you press the ENTER, or RETURN, key at the end of each line. (You do know what a typewriter is, don't you?) This action signifies that you are done with one line and ready to begin the next. In Word, however, you do not have to do this. When you set up your page margins, Word is programmed to know that when you reach the right margin, your text should automatically wrap to the next line.

Sometimes, however, you may want to end a line before you get to the right margin. In these instances, you can end a line in either of two ways. The first way is to press the ENTER key where you want the line to end. This action, which results in a ***hard return*** being entered in the document, indicates that you have reached the end of the paragraph and want to start a new one.

The other way to end a line is to press **SHIFT+ENTER**; this action results in a ***soft return,*** sometimes called a line break or a newline character, being entered in the document. A hard return is used to signify the end of a paragraph, whereas a soft return simply signifies the end of a line.

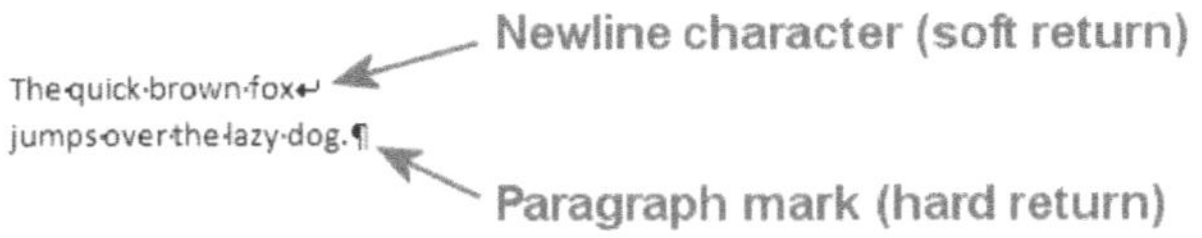

If you have changed your view options so that you can see all nonprinting characters, a hard return appears on your screen as a paragraph mark (a backward P or ***pilcrow***), and a soft return appears as a down-and-left pointing arrow.

Entering Units of Measurement in Dialog Boxes

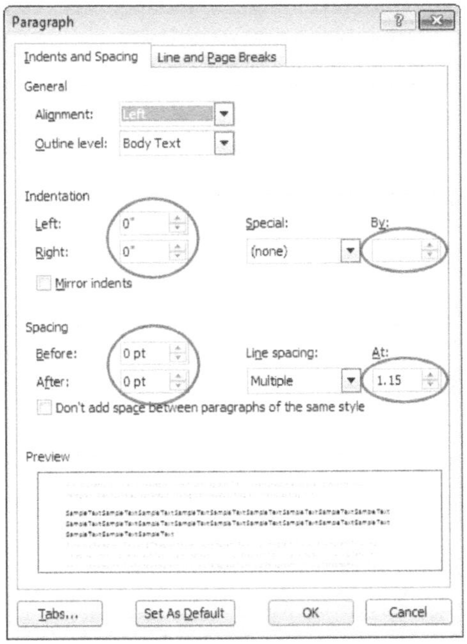

In many places in Word, you must provide measurements so that the program knows how to format elements of your document. For instance, if you display the Paragraph dialog box (which is often used for formatting paragraphs), there are numerous places where you can enter measurement values. In this and other dialog boxes, each place you can enter a measurement value is shown as a box with up and down arrows to the right of the box.

You can use the up and down arrows to change measurement values, but Word allows you to enter exact measurement values in a number of different ways. These values are typically indicated by a number followed by two letters that indicate the unit of measurement you want. Word allows you to use the following designations for units of measurement in your dialog boxes:

Abbreviation	Meaning	Example
"	Inches	.667"
cm	Centimeter	32 cm
in	Inches	.25 in
li	Lines	2.25 li
pi	Picas	8 pi
pt	Points	4 pt

Understanding Picas

A *pica* is a typographical term for a unit of measure. It is roughly equivalent to 1/6 of an inch. Picas are understood and used extensively by everyone in the publishing trade, particularly in design, typesetting, and printing. If you are familiar with picas, you can use them as a standard measurement in Word. When entering a measurement in picas, simply use the characters *pi* at the end of the measurement. For instance, to specify the space to be left after a paragraph as 2 picas, you enter **2 pi** in the After box of the Paragraph dialog box.

Understanding Point Sizes

A *point* is a typographical term for a unit of measure. It is roughly equivalent to 1/72 of an inch. Points, understood and used extensively by publishers, typographers, and designers, are most commonly used with type specifications. Word uses point sizes to specify the height of all the fonts it uses. Thus, when you use a 12-point type, you are using one that occupies a character box that is approximately 12/72 (or 1/6) of an inch high. Likewise, 72-point type uses a character box that is about 1 inch tall.

In typesetting, points are also the measurement of choice when specifying line spacing. It is not uncommon to specify type in the format 10/12, meaning 10-point type on 12-point line spacing.

If you are familiar with points, you can use them as a standard measurement in Word. When entering a measurement in points, simply use the characters *pt* at the end of the measurement.

2

Configuring Word 2010

Word allows you a great deal of latitude in controlling how it works. The process of changing how Word works or looks is generally referred to as ***configuring*** the program. Most of the tools or features of Word are configurable, as are certain portions of the user interface. The best part? What you change is entirely up to you and how you want to use Word.

This chapter focuses on a few of the more common configuration changes you can make in Word. Here you learn the answers to these questions, plus how to change how you edit words and change automatic formatting.

Changing Word's Interface

A program's ***interface*** is what you see on your screen. The interface defines how the program displays information as well as how you interact with the program. Word allows you to make many configuration changes that affect the interface it uses. In the following sections, you learn how you can control the display of different Word elements, such as the ruler, scroll bars, and the status bar. You also learn how to turn off ScreenTips and add items to the Quick Access Toolbar.

Hiding and Displaying the Ruler

The ruler appears at the top of every Word document window. The ruler is used to adjust formatting and align elements of your page. It is very helpful if you are using Word with a mouse.

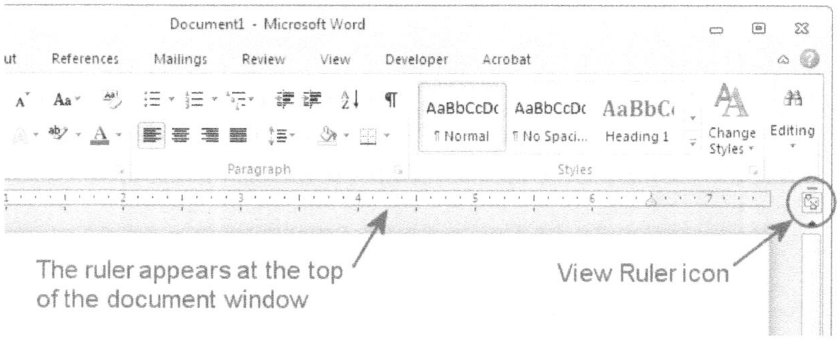

Word allows you to control whether this ruler is displayed or not. To do this, simply display the View tab of the ribbon. Note the Ruler check box in the Show group. If the check box is selected, the ruler is displayed; if it is not selected, the ruler is hidden.

There is still another easy way to display the ruler without the need to display the View tab of the ribbon. At the upper-right corner of the program window, just above the vertical scroll bar, is the View Ruler icon. Clicking this icon acts as a toggle, turning the ruler on and off, as you desire.

Controlling Scroll Bars

At the bottom and right side of the document are scroll bars that control what part of the document you are viewing. If you need more room to view a document, or if you are using Word without a mouse, you can turn off the display of the scroll bars. To control the display of the scroll bars, follow these steps:

1. Display the File tab of the ribbon and then click Options. Word displays the Word Options dialog box.
2. At the left side of the dialog box click Advanced.
3. Scroll through the options until you see the Display section.
4. Use the Show Horizontal Scroll Bar check box to turn the horizontal scroll bar on or off.
5. Use the Show Vertical Scroll Bar check box to turn the vertical scroll bar on or off.
6. Click on OK.

With the scroll bars turned off, you must use the cursor-control keys to move through the document.

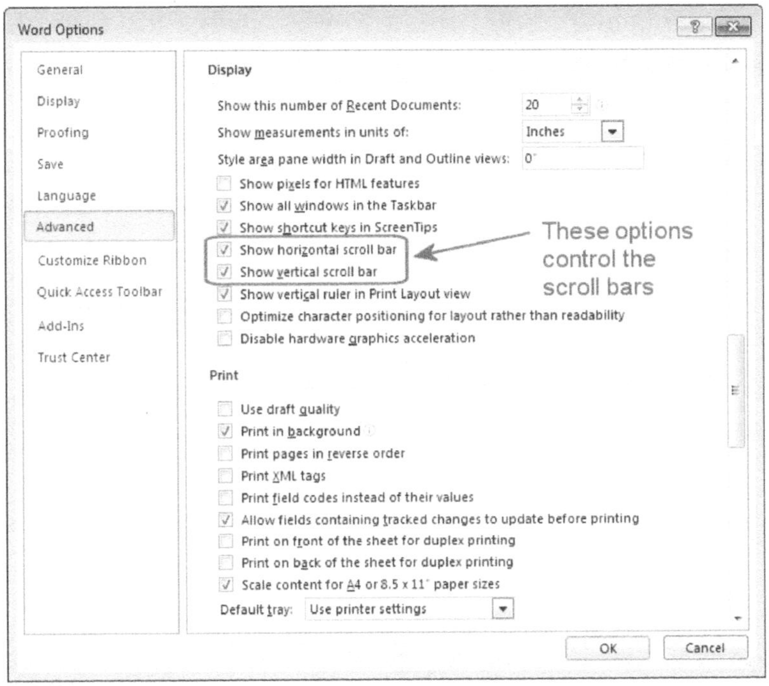

Displaying ScreenTips

As you work with the mouse in Word, you have no doubt noticed small, yellow boxes that pop up near the mouse pointer and indicate information about whatever you are pointing at. In Word, these boxes are called *ScreenTips.* By default, they are turned on, or displayed. If you find them distracting, you can turn them off by following these steps:

1. Display the File tab of the ribbon and then click Options. Word displays the Word Options dialog box.
2. Click the Advanced option at the left of the screen.
3. Scroll through the list of options until you find the Display section.
4. Make sure the Show Shortcut Keys in ScreenTips check box is cleared.
5. Click on OK.

Customizing the Quick Access Toolbar

Word allows you, within limits, to customize the tools that are available in the user interface. The primary place for customization is in controlling what appears on the Quick Access Toolbar. This is the area just above the ribbon, at the top-left portion of the screen. The Quick Access Toolbar includes, by default, three tools that allow you to save your document, undo an action, and redo an action. You can, however, change the tools that appear on the Quick Access Toolbar.

To customize the Quick Access Toolbar, follow these steps:

1. Display the File tab of the ribbon and then click Options. Word displays the Word Options dialog box.

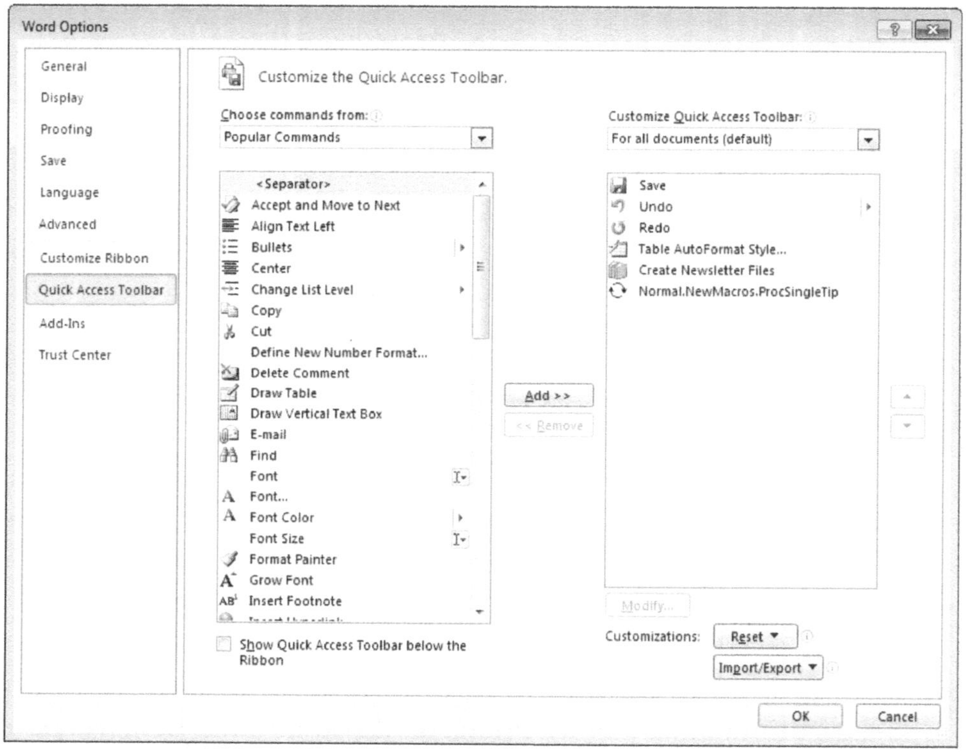

2. At the left side of the dialog box, click Quick Access Toolbar.
3. Using the Choose Commands From drop-down list, specify what group of commands you want listed in the dialog box.
4. In the list of commands at the left side of the dialog box, select the one you want to appear on the Quick Access Toolbar.
5. Click the Add button. The selected command is copied to the list at the right side of the dialog box.

6. Repeat steps 3 through 5 for any other commands you want on the toolbar.
7. If you want to remove any commands from the Quick Access Toolbar, select the command at the right side of the dialog box and click Remove.
8. When you are satisfied with the list of commands, click OK.

Your newly customized Quick Access Toolbar should be visible and usable at the top-left of your screen.

Changing What You See

So far in this chapter, you have learned how to change the interface used by Word. You can also make a couple of changes that affect how your document is displayed on the screen. In the following sections, you learn how to view all your text when working with wide documents as well as how to see where spaces appear in a document.

Horizontally Viewing All Your Text

Normally, Word displays text on the screen as it will appear on a hard-copy printout. This feature is often referred to as WYSIWYG (pronounced "whizzy-wig"), which means "what you see is what you get." At times, however, viewing documents this way can be bothersome—for instance, if you are working on a very wide document or using a computer set to a low video resolution. In these instances, the only way to see all your text horizontally is to scroll left and right—which can get to be tedious.

If you get tired of scrolling left and right, you can try the following:

1. Display the File tab of the ribbon and then click Options. Word displays the Word Options dialog box.
2. Click Advanced at the left of the dialog box.
3. Scroll through the available options until you see the Show Document Content section.

4. Make sure the Show Text Wrapped Within the Document Window check box is cleared.
5. Click on OK.

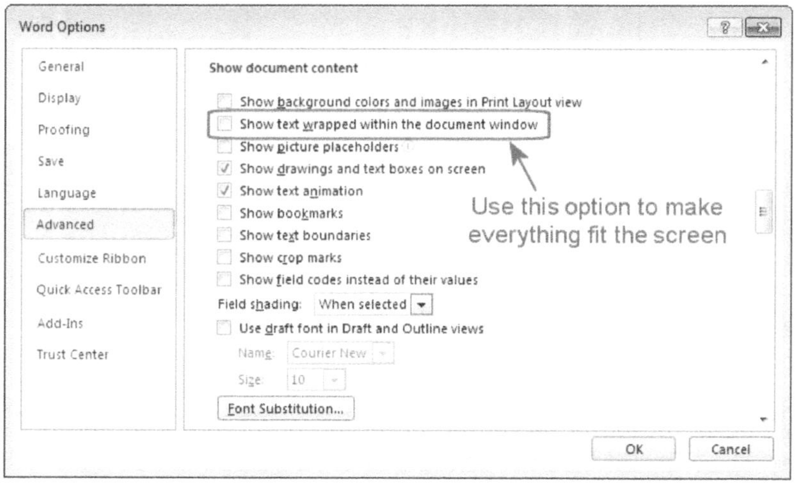

This check box essentially turns off most of the WYSIWYG features of Word. Instead, none of your document lines extends beyond the rightmost edge of your screen. When you later want to see what your document will look like on a printout, simply repeat these steps to clear the check box or use the Print Preview feature of Word.

You should note that this setting only works if you are using either the Normal or Outline views within Word. If you are using any other view, the setting is ignored.

Viewing Your Spaces

Have you ever worked on a document that included multiple fonts and the printout didn't quite match what you wanted? Or, maybe there was more space between words or sentences than what seemed right and you didn't quite know what the problem was?

I have found that the culprit is often me. I may have placed extra spaces between words or sentences in such a manner that the text looks fine on screen, but doesn't quite fit the bill when it's finally printed. The solution, of course, is to remove the extra spaces and make sure that there are a consistent number of them between both words and sentences. The problem with this solution is that it is often difficult to see how many spaces your document really has.

Fortunately, Word allows you to easily see where spaces occur in your documents. You do this by telling Word to display nonprinting characters, as discussed in Chapter 1, "Getting Acquainted with Word 2010." After they're displayed, all your spaces appear as small, visible dots. You can easily tell where two visible dots appear, as opposed to two invisible spaces. Delete extra spaces as you would delete any other characters, and you will have your document cleaned up in no time.

Handling Changes That Affect Editing

Most of the time you work with Word, you are editing documents. It seems that changes to wording go on forever. (On some documents or projects, the changes *do* go on forever.) The tips in the following sections don't tell you how to do editing; that task is best left to Chapter 3, "Creating and Editing Documents." Instead, you learn here how to configure Word to make changes in how editing is done.

Automatically Selecting Words

When you are editing your document, Word often tries to guess what you are doing and then completes your task for you. This helpfulness can be aggravating at times. Once such instance is when you are trying to use the mouse to select text from the middle of one word to the middle of another. Whenever you click inside a word and then start dragging the mouse, Word assumes that you want to make a selection. As soon as you move the mouse pointer outside the word, the program assumes that you want to select the entire first word in which you originally clicked.

If you want to ensure that Word doesn't make that assumption, follow these steps to turn off the offending feature:

1. Display the File tab of the ribbon and then click Options. Word displays the Word Options dialog box.
2. Click Advanced at the left side of the screen.

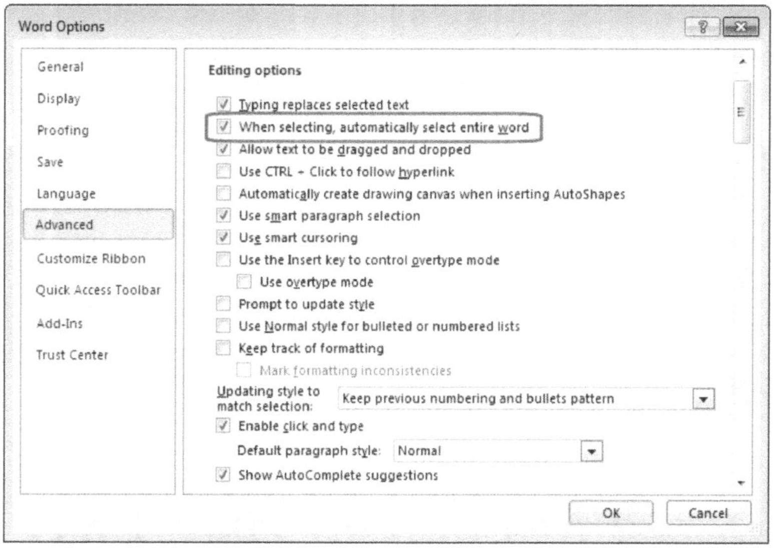

3. In the Editing Options area, make sure the When Selecting, Automatically Select Entire Word check box is cleared.
4. Click on OK.

Using the Insert Key to Insert Text

When you first install Word, the **Insert** key is used to turn overtype mode on and off. (*Overtype mode* means that what you type "types over" what is already in the document; it is discussed more fully in Chapter 3, "Creating and Editing Documents.") Some people find this action to be awkward, particularly if they are familiar with working with other word processors that use the **Insert** key differently. In some word processors, the **Delete** key is used to delete text and the **Insert** key is used to place the text at the current cursor position.

Word can work in this way as well. You can set the **Insert** key to function the same as the **Ctrl+V** key combination—to insert the contents of the Clipboard. To control the way the **Insert** key works, follow these steps:

1. Display the File tab of the ribbon and then click Options. Word displays the Word Options dialog box.

2. At the left side of the dialog box click Advanced.
3. Scroll down until you see the Cut, Copy, and Paste section.
4. Make sure the Use the Insert Key for Paste check box is selected.
5. Click OK.

Changing the Default Measurement Units

Word, as you already know, allows you to modify your work environment. One change you can make involves the default units of measurement used in dialog boxes and on the ribbon. By default, Word uses inches, but you can customize the program so that it uses centimeters, picas, or points. To change the default measurement units, follow these steps:

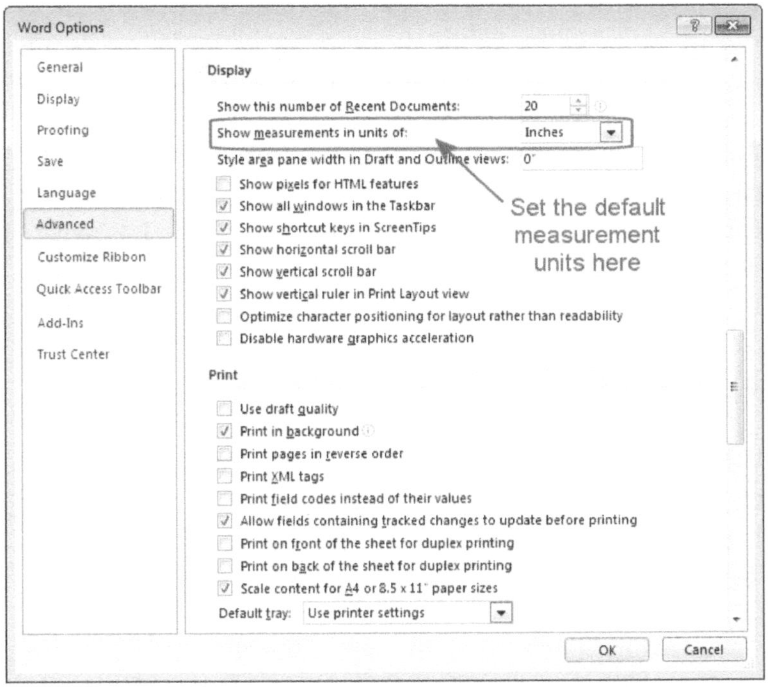

1. Display the File tab of the ribbon and then click Options. Word displays the Word Options dialog box.
2. Click Advanced at the left side of the dialog box.
3. Scroll through the available options until you see the Display section.
4. Using the Show Measurements in Units Of drop-down list, select the units of measurement you want used by default.
5. Click on OK.

Controlling Automatic Formatting

In Word, editing and formatting are two separate tasks. You learn how to format your documents in Chapter 5, "Formatting Your Document." As you discover in that chapter, Word can handle some formatting needs for you automatically. In the following sections, you learn how you can change the way in which Word formats, just by making a few configuration changes.

Turning Off Automatic Lists

Word includes a feature that can automatically format special lists in your documents. These lists include both numbered and bulleted lists. When you enter some text that Word thinks should be a list and then press **ENTER**, Word formats the paragraph with a hanging indent and places an appropriate number or bullet at the beginning of it. In addition, Word assumes that the next paragraph will be part of the same list.

For instance, you might type a number or letter and then a period and then press the **SPACE BAR** or **TAB** key. As soon as you type the space or tab, Word formats the paragraph as a numbered list. Likewise, if you type an asterisk, press the **SPACE BAR** or the **TAB** key, Word formats the paragraph as a bulleted list.

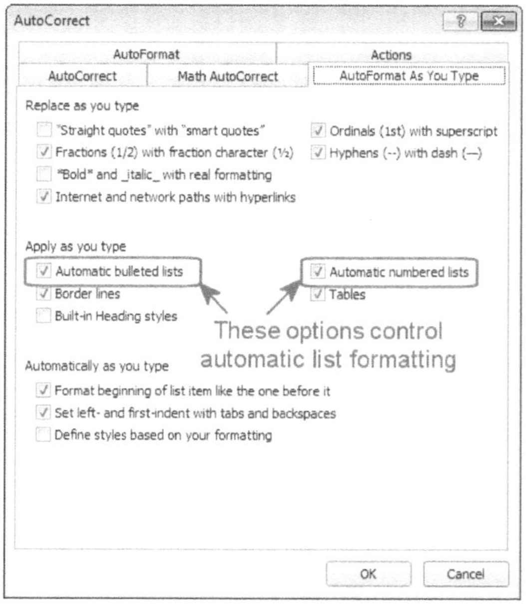

If you don't want the paragraph to be formatted automatically, you can cancel the formatting by immediately pressing **CTRL+Z**. If the automatic formatting of lists bothers you, you can follow these steps to turn it off:

1. Display the File tab of the ribbon and then click Options. Word displays the Word Options dialog box.

2. Click Proofing at the left side of the dialog box, and then click AutoCorrect Options. Word displays the AutoCorrect dialog box.
3. Make sure the AutoFormat As You Type tab is displayed.
4. Clear the Automatic Bulleted Lists and Automatic Numbered Lists check boxes.
5. Click on OK to close the AutoCorrect dialog box.
6. Click on Cancel to close the Word Options dialog box.

Turning Off Automatic Capitalization

As you are typing away on a document, you may have noticed that periodically Word second-guesses what you are doing and capitalizes words for you. In general, Word does this when it thinks you are starting a new sentence. If you find yourself undoing Word's decisions on this issue quite a bit, you can turn off the feature by following these steps:

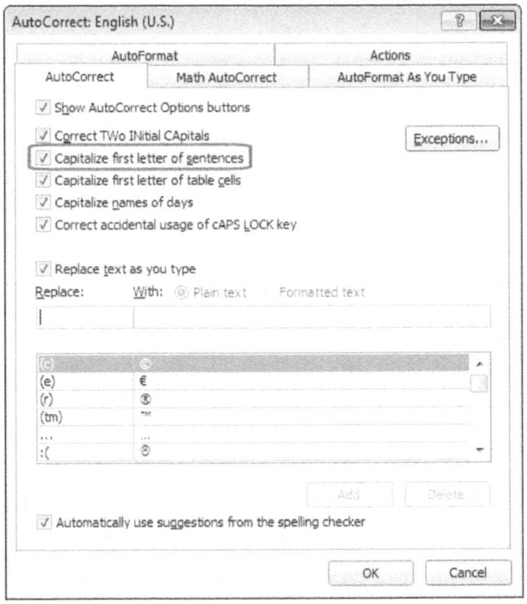

1. Display the File tab of the ribbon and then click Options. Word displays the Word Options dialog box.
2. Click Proofing at the left side of the dialog box, and then click AutoCorrect Options. Word displays the AutoCorrect dialog box.
2. Make sure the AutoCorrect tab is displayed.
3. Clear the Capitalize First Letter of Sentences check box.
4. Click OK to dismiss the AutoCorrect dialog box.
5. Click on Cancel to dismiss the Word Options dialog box.

Turning Off Capital Corrections

Word tries its level best to be a good guesser at what you are trying to do. For instance, as you are typing along, if you type a word where the first two letters are uppercase and the next letter is lowercase, Word figures that you just have slow fingers and didn't release the **Shift** key in time to make the second letter lowercase. So, it dutifully changes the second letter to lowercase to help you out. For instance, the word **PLace** becomes **Place.**

In some situations, this behavior can be bothersome, however. For instance, the first two letters of your company name or product name may always be capitalized, such as INtec or MYphone. In these cases, Word also tries to do its magic and change the capitalization, which can cause no end to proofreading passes and related problems.

One solution to this problem is to turn off the correction that Word does to your words (at least for this particular capitalization issue). The way you do that is as follows:

1. Display the File tab of the ribbon and then click Options. Word displays the Word Options dialog box.
2. Click Proofing at the left side of the dialog box, and then click AutoCorrect Options. Word displays the AutoCorrect dialog box.
3. Make sure the AutoCorrect tab is displayed.
4. Clear the Correct TWo INitial Capitals check box.
5. Click OK to dismiss the AutoCorrect dialog box.
6. Click on Cancel to close the Word Options dialog box.

Automatically Capitalizing Day Names

Every time you type a word in Word, the program does quite a bit of behind-the-scenes work to check out what you typed and, if necessary, make changes to it. One thing Word can automatically check and correct is the capitalization of days of the week. For instance, if you type the word ***wednesday***, Word automatically changes it to ***Wednesday***. To control this behavior in Word, follow these steps:

1. Display the File tab of the ribbon and then click Options. Word displays the Word Options dialog box.
2. Click Proofing at the left side of the dialog box, and then click AutoCorrect Options. Word displays the AutoCorrect dialog box.
3. Make sure the AutoCorrect tab is selected.
4. Select the Capitalize Names of Days check box if you want Word to automatically capitalize for you.
4. Clear the Capitalize Names of Days check box if you don't want Word to make automatic changes.
5. Click OK to dismiss the AutoCorrect dialog box.
6. Click on Cancel to close the Word Options dialog box.

3

Creating and Editing Documents

The main role of any word processing program is to allow you to create and edit documents. Gone, thankfully, are the days of the typewriter, when you needed to throw away previous manuscript versions and start from scratch whenever you wanted to make extensive changes.

Word provides a rich set of editing tools. After you start typing, you use these tools to slice, dice, and fricassee your text into just the texture and flavor you want. In the following sections you learn about special characters you can add to your text, how to navigate through your document, and how to select text. You learn how to make changes to your text and copy or move text. Finally, you learn how to work with headers, footers, and page numbers.

Understanding Breaks and Other Special Characters

Normally, Word treats your text as one continuous stream of information. Letters, numbers, and punctuation are combined in the text stream to create the phrases, sentences, and paragraphs of your document. As Word displays this information on the page, it automatically flows from the bottom of one page to the top of the next.

At times, you may want to interrupt the continuous stream of your text. In the following sections, you learn how you can do this by using breaks. You also learn how you can use hyphens, dashes, special spaces, and other symbols to modify and enhance your text.

Breaking Your Text

As you are typing in Word, notice that the program automatically keeps track of where you are. When your document no longer fits on a single page, Word automatically inserts a page break (indicated by a thin dashed line when viewing your document in Normal view) and continues your text on the next page.

At times, however, you may want to insert a break in your text manually. Word supports these types of breaks:

- **Page break.** Causes the following text to begin at the top of the next page.
- **Column break.** Causes the following text to begin at the top of the next column. (If only one column is in your page layout, the column break functions the same as a page break.)
- **Section break.** Causes the following text to begin a new section of your document. Where the actual break occurs depends on the type of section break you use. Four different types of section breaks can be inserted in your document.
- **Text-wrapping break.** Closely akin to a line break (SHIFT+ENTER). A text-wrapping break breaks a line of text and moves the text to the next line. This type of break is intended for use with text that wraps around graphics.

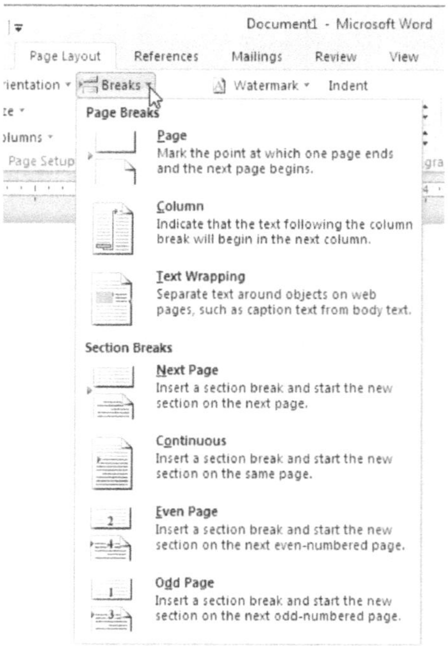

To insert a break in your document, follow these steps:

1. Position the insertion point at the position where you want to insert the break.
2. Display the Page Layout tab of the ribbon.
3. Click the Breaks tool, in the Page Setup group. Word displays a list of breaks you can insert.
4. Click the type of break you want to insert.

Removing a Break

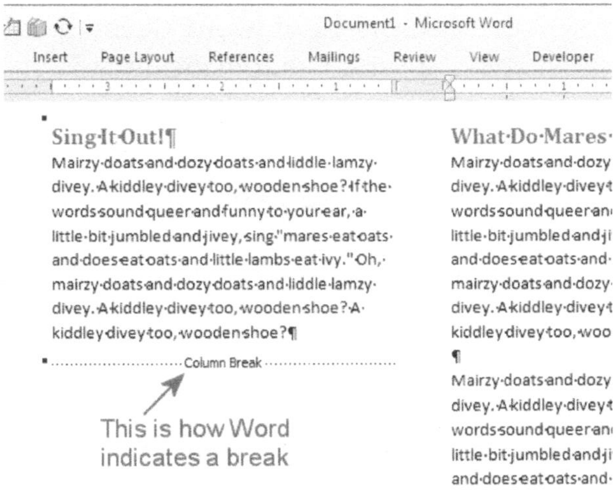

Breaks are used quite often in a document to disrupt the normal contiguous flow of text. As long as you have non-printing characters displayed in your document—as discussed in Chapter 1, "Getting Acquainted with Word 2010"—then Word displays them as lines extending the full width of your column or page. The type of break is denoted by text in the middle of the line. For instance, column breaks appear as a thin dotted line, with the text ***Column Break*** in the middle of the dotted line.

Breaks are treated like any other characters in Word. If you want to delete a break, simply position the insertion point on the break and use any of the normal editing keys to delete it. Perhaps the quickest way to remove a break is to position the insertion point on the break and press the **Del** key.

Understanding Hyphens and Dashes

Word supports the use of both hyphens and dashes. Actually, it supports three types of hyphens and two types of dashes. You should understand how Word handles each of these characters because they can affect the appearance of your document:

- **Regular hyphens.** These are created by simply typing the hyphen key. This is the key that is to the right of the zero key on the keyboard. It is sometimes mistakenly called a dash key. Regular hyphens are used to create compound words, such as "mix-up," or to indicate a minus sign in an equation. If a compound word appears near the end of a line, the second word will be displayed on the next line, if necessary, with the first word and hyphen remaining on the previous line.
- **Optional hyphens.** These are created by pressing **Ctrl+-** (**Ctrl** and the hyphen key). Optional hyphens are typically used in the middle of a word, between syllables, to indicate where a word should be broken between lines, if Word deems it necessary. Optional hyphens are the type inserted automatically when you use the Hyphenation tool in Word. The optional hyphen does not appear on any printout unless it is actually used at the end of a line.
- **Non-breaking hyphens.** These are created by pressing **Ctrl+Shift+-** (**Ctrl+Shift** and the hyphen key). Non-breaking hyphens are used in compound words to indicate that both words and the hyphen should be treated as a single word when Word is forming lines. In this case, the compound word will never be broken over two lines. It is also helpful to use non-breaking hyphens in phone numbers.
- **En dash.** An en dash is a typographic dash that is as wide as a lowercase "n" character. These dashes are typically used to denote ranges of numbers, as in 3–7. You create an en dash by pressing **Ctrl** and the minus sign on the numeric keypad. You can also create it by holding down the **Alt** key as you type 0150 on the numeric keypad. If necessary, Word will break a line right after the en dash, not before it. In other words, the en dash always stays with the characters immediately preceding it.
- **Em dash.** An em dash is a typographic dash that is supposed to be as wide as a lowercase "m" character. In Word, however, the em dash is twice as wide as the en dash. (The width of the em-dash can vary from font to font.) Em dashes are used in creating breaks in sentences, between two separate thoughts. Word will substitute an em dash automatically as you are typing if you type a word, two hyphens in a row, and another word. You can also explicitly enter an em dash if you press **Ctrl+Alt** and the minus sign on the numeric keypad, or you can enter one by

holding down the Alt key as you type 0151 on the numeric keypad. If necessary, Word will break a line right after the em dash, not before it. The em dash always stays with the word immediately before it.

The foregoing items describe the behavior and purpose of each of the hyphens and dashes used in Word. If you are using a dash or hyphen and it does not behave as indicated here, you may be using the wrong type. For instance, if a word does not break as you expect between two lines, you may be using a nonbreaking hyphen instead of one of the other types.

Inserting a Nonbreaking Space

When Word calculates line width and wraps text to the next line, it tries to break the line at either a space or a hyphen. Sometimes, however, you may not want Word to break a line at a certain space. For instance, you may want to make sure that two adjacent words appear on the same line as each other. The answer is to use nonbreaking spaces instead of regular spaces when you don't want Word to break a line at a certain space. To do this, hold down the Ctrl and Shift keys as you press the Space Bar. Word then does not break the line at that point.

An alternative way of inserting a nonbreaking space is to follow these steps:

1. Display the Symbol dialog box by displaying the Insert tab of the ribbon, clicking Symbol, and then More Symbols.
2. Click on the Special Characters tab.
3. Highlight the Nonbreaking Space character.
4. Click on Insert.
5. Close the dialog box by clicking on Cancel.

Inserting a Special Symbol

In typography, many special symbols are defined in fonts. An example of special symbols are the bullets used at the front of bulleted lists. Word allows you to quickly and easily insert a special symbol into your document. To do so, follow these instructions:

1. Position the insertion point where you want the special symbol inserted.
2. Display the Symbol dialog box by displaying the Insert tab of the ribbon, clicking Symbol, and then More Symbols.

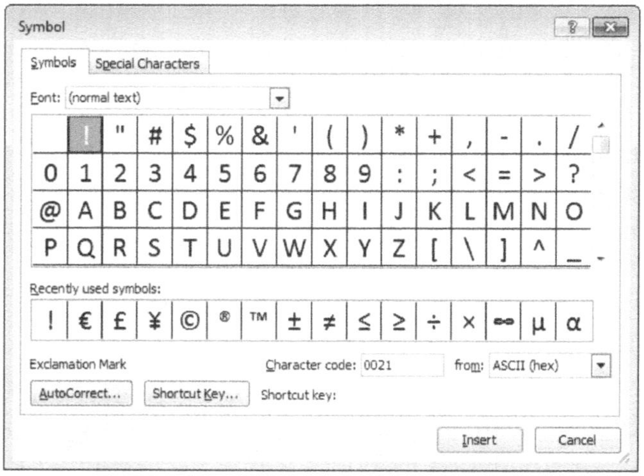

3. Make sure the Symbols tab is selected. (It should be selected by default when you first call up the dialog box.)
4. Since the symbols shown are nothing more than alternative typefaces, you can pick a different symbol typeface by changing the Font drop-down list.
5. Select a symbol from among those displayed.
6. Click on Insert.
7. Repeat steps 4 through 6 to insert other symbols, if desired.
8. Click on Cancel to close the dialog box.

Navigating Within Your Document

You sit down one morning to write a quick letter to your aunt, and, before you know it, you've written the Great American Novel. When faced with documents that contain dozens or hundreds of pages, you need to know how to navigate through the document. The following sections teach you just the skills you need.

Picking Up Where You Left Off

If you find yourself working on the same document in multiple editing sessions, you can use the information in this section to get started faster. If you open a document you were editing when you last used Word, you can use the **Shift+F5** keyboard combination to quickly jump to where you were last working in the document. This is also a good way to get back to what you were editing after browsing through a document.

You should remember an important caveat when trying to apply this tip: The **Shift+F5** shortcut takes you back to where you made your last edit, not to where you last had the insertion point. For instance, if you are halfway through your document and you type a word (or change some characters), that counts as an edit. Suppose that you then press the **Page Down** key a few times to get to the bottom of the document, and then you save and close the document. When you reopen the document and press **Shift+F5**, Word takes you to where you made the edits (halfway through the document), not to where you were when you saved the document (at the end).

Using Go To to Move Around in a Document

You can use the Go To function to jump to just about any place you desire in your document. All you need to do is press the **F5** key and Word displays the Go To tab of the Find and Replace dialog box.

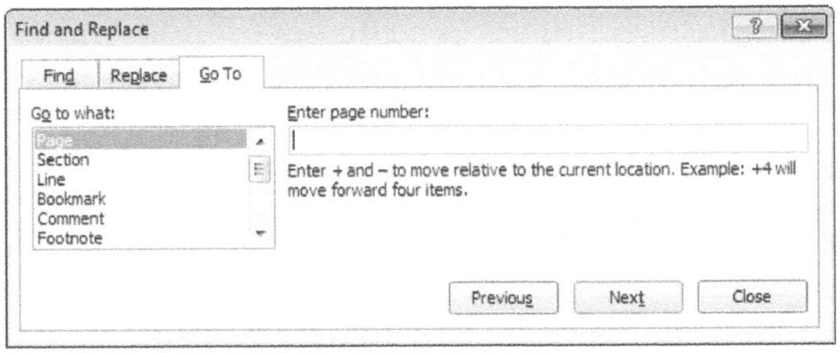

At the left side of the dialog box is a list of different things you can "go to." You can select from 13 different items: Page, Section, Line, Bookmark, Comment, Footnote, Endnote, Field, Table, Graphic, Equation, Object, and Heading.

The general idea when using the Go To function is to select (from the item list) what you want to go to and then, in the box on the right side of the dialog box, enter a specific value that indicates how you want to jump to that item. When you specify a value in the text box, you can do it in several ways:

- **Absolute value.** If you enter a number, Word assumes you want to go to that occurrence of whatever item you selected on the left side of the dialog box. For instance, if you enter a 5 and select Page, Word assumes that you want to jump to page 5 in your document.

- **Relative value.** If you put a minus sign or a plus sign in front of a number, Word jumps a relative number of items from the current location in the document. For instance, if you click on Line and then enter –2 in the text box, Word jumps backward two lines in the document.
- **Percentage.** You can enter a percentage in the text box, which causes Word to jump that percentage of objects from the beginning of the document. This option is typically used with the Page item. If you enter, for instance, 75% in the text box (or %75; Word doesn't care), Word jumps 75 percent of the way through the document.
- **Name.** If the item you pick in the list of items relies on names, you can select the named item. For instance, if you pick Bookmark, you can specify the name of a bookmark to which you want to jump. (You can enter names only with Bookmark, Field, and Object items.)

Word also allows you to leave empty the text box on the right side of the dialog box. If you do, Word assumes you want to go to the next occurrence of whatever you pick in the items list.

When you have specified the item you want to go to, simply click on the Next button (if you didn't specify anything in the text box) or the Go To button (if you did specify something in the text box).

Jumping Between Columns

If you are working with multiple columns in your document, you may need to jump from column to column at times. The standard way to do this (using the keyboard) is to use the **Alt** key in conjunction with the up and down arrow keys. If you press **Alt+Down Arrow**, the insertion point is moved to the top of the next column. If you press **Alt+Up Arrow**, the insertion point goes to the top of the previous column.

This feature works only if you are viewing your document in Print Layout view. Plus, if only a partial number of columns is visible on the screen, this movement method cycles only between the visible columns. For instance, you may have your document formatted for three columns. If you are viewing the last page of the document, one, two, or three columns could be visible. (You might not have enough text to fill up three columns on that last page.) In this case, you cycle only between the columns on the screen.

Selecting Text

The standard method of performing an action in Word is to first select the text you want to affect and then issue the command that performs the action. Thus, selecting text is integral to editing text. In this section, you learn the different ways you can select your text, in preparation for the next section, which covers the actual editing.

Quickly Selecting Text

You already know that you can use the mouse to select text by holding down the left mouse button and dragging the mouse to make a selection. You can use an even quicker way to select text using the mouse, however:

1. Position the insertion point where you want the selection to begin.
2. Move the mouse cursor so that it points to where you want the selection to end.
3. Hold down the **Shift** key as you click once on the left mouse button.

Selecting a Word

If you are using a mouse, Word provides a shortcut to select an entire word. To do this, follow these two steps:

1. Using the mouse pointer, point to the word you want to select.
2. Double-click on the mouse.

If you want to select additional words, hold down the mouse button after the second click and drag the mouse across them. Word adds one word at a time to your selection. After the words are selected, you can do any other editing function on those words.

Selecting a Sentence

You already know how to select words using the mouse. You may not know, however, that you can use the mouse to select a sentence.

To select a sentence using the mouse, simply hold down the **Ctrl** key as you click anywhere within the sentence. The entire sentence, along with any trailing spaces, is selected. Note that Word does get a little confused if the sentence contains a period for an abbreviation (such as in Mr. or Mrs. or Dr.). If you have such a sentence, simply continue to hold down the **Ctrl** key as you move the mouse past the abbreviation. The rest of the sentence is then selected.

Selecting a Line of Text

If you are new to Word, coming from using a different type of word processor, you probably noticed that Word has no command to delete a line of text. You can do this quickly by moving the mouse pointer to the left of the line of text you want to select. Don't move the mouse pointer into the style area (if it is displayed), but just to the left of the first character in the line. The mouse pointer turns into an arrow pointing up and right. Click once on the left mouse button, and the entire line of text is selected. You then can do any other editing function on that line of text, including deleting it (by pressing **Del** or **Ctrl+X**).

Selecting an Entire Paragraph

If you are using the mouse, Word provides several quick ways to select an entire paragraph. First, you can simply triple-click anywhere within the paragraph. This method is perhaps the fastest and most common way of selecting an entire paragraph.

Second, you can move the mouse pointer to the left of the first character in any line of the paragraph you want to select. The mouse pointer turns into an arrow pointing up and right. Double-click the mouse, and the paragraph is selected.

Finally, if you are using the style area, you can also move the mouse pointer into the style area to the left of the paragraph you want to select. Click once on the left mouse button, and the entire paragraph is selected.

If you like to use the keyboard instead of the mouse, you can select the current paragraph by pressing **Ctrl+Up Arrow** (which moves the insertion point to the beginning of the paragraph) and then pressing **Ctrl+Shift+Down Arrow**.

Using Extend Mode

If you press the **F8** key, you enable what Word refers to as *extend mode.* It is the keyboard method of anchoring the cursor in preparation for making a selection. When you move the cursor using the cursor-control keys, the size of the text selection changes. You turn off extend mode by pressing the **Esc** key.

When you press the **F8** key multiple times, you can select different parts of your text. For instance, when you press **F8** once, you simply turn on extend mode. Press it a second time, and the current word is selected. A third time results in the current sentence being selected. A fourth time selects the current paragraph, and a fifth time selects the whole document.

Extending a Selection

As you learned in the preceding section, you can turn on extend mode (which allows you to make selections) by pressing the **F8** key. Some people may wonder why anyone would bother to use **F8** to make selections when simply holding down the **Shift** key or selecting with the mouse is just as easy, if not easier.

The answer is that using **F8** enables a handy feature that is not available in other selection methods. If you press **F8** to turn on extend mode, you can then press any other key and Word automatically extends the selection up to the next occurrence of that character. For instance, if you want to select everything from the current position of the insertion point through the next comma, simply press **F8** and then type a comma—the selection is automatically made. Type another comma, and the selection is extended to the next comma.

You can press any other key you want, except for **F8** or **Esc**. Either of these cancels extend mode. If there is no subsequent occurrence of the key you press, Word does not extend the selection.

Selecting a Text Block

When working with some forms of data in Word (particularly tabular columns of information), it is often helpful to select nonsequential text in a block. For instance, you might want to select the tenth through thirtieth characters on each of five lines, ignoring everything else. Word makes this easy to do, using either the keyboard or the mouse. If you want to select a block of text using the keyboard, follow these steps:

1. Position the insertion point at the position that defines the upper-left corner of the block.
2. Press **Ctrl+Shift+F8**.
3. Use the cursor control keys to extend the block to include all the text desired.

If you would rather use the mouse to block the text, you can do so by simply holding down the **Alt** key as you make your selection.

After the text block is selected, you can take any action you want. For instance, if you press the **Del** key, the block of text is deleted.

Selecting the Entire Document with the Mouse

If you need to select the entire document, you can do so with the mouse by following these steps:

1. Move the mouse pointer to the left of the paragraphs in your document. The pointer changes to an arrow pointing up and to the right.
2. Hold down the **Ctrl** key.
3. Click the left mouse button.

Another way to do this is to simply triple-click the left mouse button after step 1. If you prefer to not use the mouse, you can also select the entire document by pressing **Ctrl+A**.

Editing Your Text

It's time to edit! As mentioned earlier in this chapter, Word provides a rich array of tools you can use to edit your text. The following sections describe the ways you can use those tools so that your text appears exactly as you want it. By the time you are done with this section, you will know how to use Word's Click and Type feature, use the Clipboard, replace text selections, put overtype mode to work, and repeat your typing on demand.

Understanding Click and Type

Word includes a feature known as ***Click and Type***. This feature means that when you are working in Print Layout view or Web Layout view, you can double-click in any open area of your document (where there is not text) and begin typing right away.

Normally, you begin typing at the left side of the screen. If you later want to adjust your text to the right margin or center it on the screen, you do so by using the tools on the various ribbon tabs. Click and Type, however, allows you to quickly format and enter text at the same time. The result is faster editing and formatting.

You can tell if Click and Type is active by how the mouse pointer behaves on-screen. If the mouse pointer, when moving within the document window, looks like a simple I-beam insertion point, Click and Type is not turned on. If, instead, the mouse pointer looks like an I-beam with some horizontal lines nearby, Click and Type is active and ready. These lines indicate the formatting of the text you can insert. There are four possibilities:

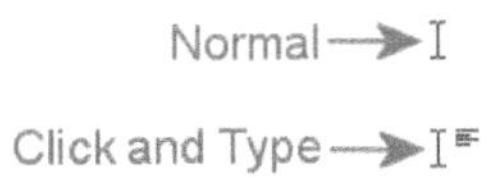

- **Left aligned.** If the horizontal lines are near the upper-right side of the I-beam, it indicates that double-clicking results in a left-aligned paragraph where you click.
- **Left aligned, first line indent.** If the horizontal lines are near the upper-right side of the I-beam and a very small arrow is on the left side of the first horizontal line, double-clicking results in a left-aligned paragraph where you click, with the first line of the paragraph indented.

- **Centered.** If the horizontal lines are directly beneath the I-beam, you can enter a centered paragraph by double-clicking.
- **Right aligned.** If the horizontal lines are near the upper-left side of the I-beam, it indicates that double-clicking adds a right-aligned paragraph where you click.

Remember that Click and Type works only if you are viewing your document in Print Layout view or in Web Layout view.

Understanding the Default Click and Type Paragraph Style

Word uses ***styles*** to define the way paragraphs and characters should appear. (You discover more about styles in Chapter 6, "Introducing Styles and Templates.") Normally, when you enter a paragraph by using the Click and Type method, the paragraph is essentially formatted using the Normal style. Granted, the basic formatting of the style has some changes (such as right alignment, center alignment, or indentation), but it is nonetheless the Normal style.

You can control what style Word uses as the base style for any paragraph you enter using Click and Type. To do this, follow these steps:

1. Display the File tab of the ribbon and then click Options. Word displays the Word Options dialog box.
2. At the left side of the dialog box, click Advanced.

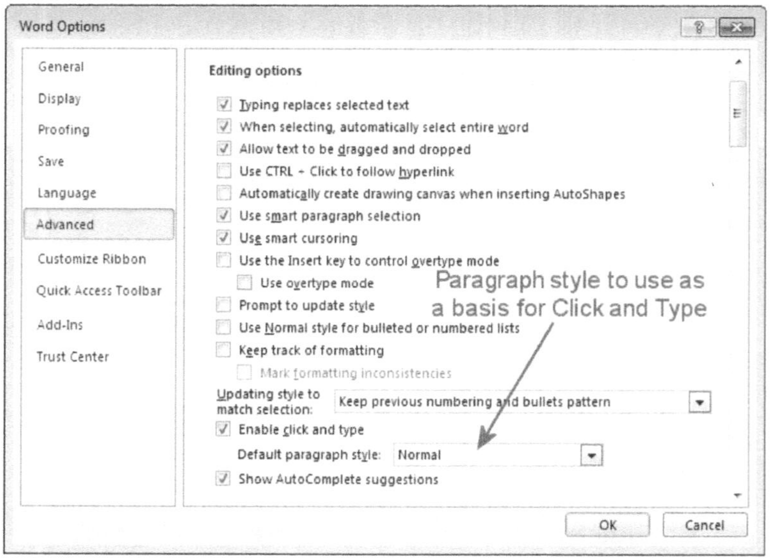

3. In the Editing Options area of the dialog box, use the Default Paragraph Style drop-down list to specify a style to use for your Click and Type paragraphs.
4. Click on OK.

Turning Off Click and Type

If you don't like or don't use the Click and Type feature in Word, you may want to turn it off. You can do this by following these steps:

1. Display the File tab of the ribbon and then click Options. Word displays the Word Options dialog box.
2. At the left side of the dialog box, click Advanced.
3. In the Editing Options area of the dialog box, make sure the Enable Click and Type check box is cleared.
4. Click on OK.

Understanding the Clipboard

Any long-time Windows user knows that the ***Clipboard*** is the place where information is temporarily stored when you are cutting or copying. The Clipboard is a feature of Windows, but it is supported completely by Word. (There is also an Office Clipboard; more on that in a moment.) You place information in the Clipboard by first selecting it and then choose Copy or Cut from the tools available on the Home tab of the ribbon. You can also press **Ctrl+C** (to copy) or **Ctrl+X** (to cut).

When information is stored in the Clipboard, Windows also keeps track of the type of information stored. For instance, if you copy some text to the Clipboard, Windows tracks the source of the text. The reason for tracking the source is so the information can be correctly pasted into a different program. This means that if you copy cells from an Excel worksheet, you can paste them into Word in a variety of formats.

If you have something in the Clipboard, you can either paste it into Word by clicking the Paste tool on the Home tab of the ribbon or by pressing **Ctrl+V**. If you want more control over how something is pasted from the Clipboard, use the Paste Special tool on the Home tab of the ribbon.

Microsoft Office (which includes Word) also has an expanded version of the Clipboard. The Office Clipboard interacts with the regular Windows Clipboard in that the latest item copied to the Clipboard is in both the Windows and Office clipboards. The Office Clipboard, however, will keep track of many more than a single cut or copy. You can display the contents of the Office Clipboard by displaying the Home tab of the ribbon and then clicking the small icon at the bottom-right of the Clipboard group. Word displays the Clipboard pane at the left side of the document.

Clicking any of the items in the Clipboard pane pastes it wherever the insertion point is in the document. The Options drop down list at the bottom of the Clipboard pane sets options of when to automatically display the Clipboard pane and Clipboard indicators displayed at the system taskbar.

Understanding the Differences Between Deleting, Clearing, and Cutting

In most books you read about Word, and even in the Word documentation, you see the terms *delete, clear,* and *cut,* each of which refers to an action you can do with text. These words are almost interchangeable, but they have subtle differences:

- **Delete** means to remove text or another element from your document by pressing the **Del** key. This task is analogous to clearing.
- **Clear** means to remove text or another element from your document without saving it in the Clipboard. This task is typically done with the **Del** or **Backspace** key.
- **Cut** means to remove text or another element from your document and save it in the Clipboard. This task is accomplished by using the **Ctrl+X** shortcut keys or by choosing Cut from the Home tab of the ribbon.

Clearing and cutting also have specific meanings if you are working with tables in Word. If you select cells in a table and clear them (by pressing **Del** or **Backspace**), the information in the tables is deleted, but the table itself remains intact. If you select an entire row or column and delete (by pressing **Ctrl+X**), the row or column is removed from the table.

Replacing Text Selections

If you have selected some text and then press any key except a cursor-control key, the selected text is, by default, replaced with what you type. This replacement can be annoying, especially if this behavior does not match what you are familiar with in another word processor. To disable this aspect of Word, follow these steps:

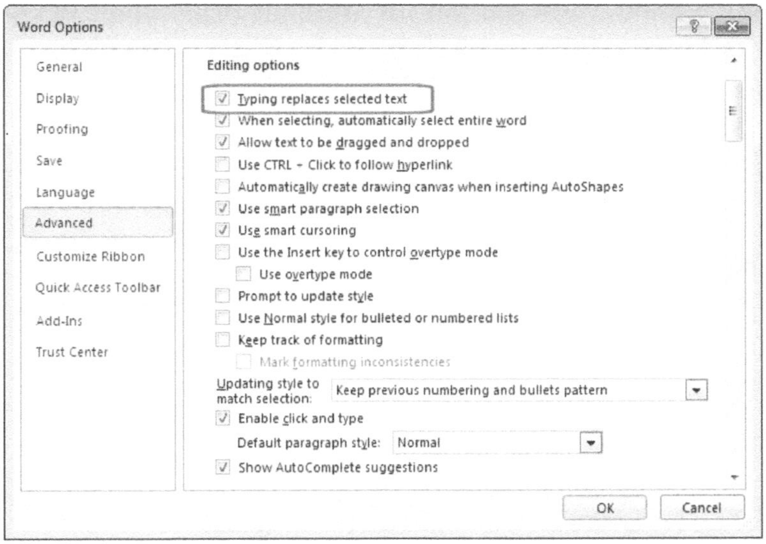

1. Display the File tab of the ribbon and then click Options. Word displays the Word Options dialog box.
2. Click Advanced at the left side of the dialog box.
3. Click on Typing Replaces Selected Text. (This should be the first option in the dialog box.) If the check box is selected, then what you type replaces what you have selected; if it is not selected then what you type is inserted before what you have selected.
4. Click on OK.

Using Overtype Mode

Overtype mode is an editing mode in which everything you type replaces something else in your document. When overtype mode is active and you type a letter, it replaces the letter to the right of the insertion point. When overtype mode is not active, your text is inserted where the insertion point is located.

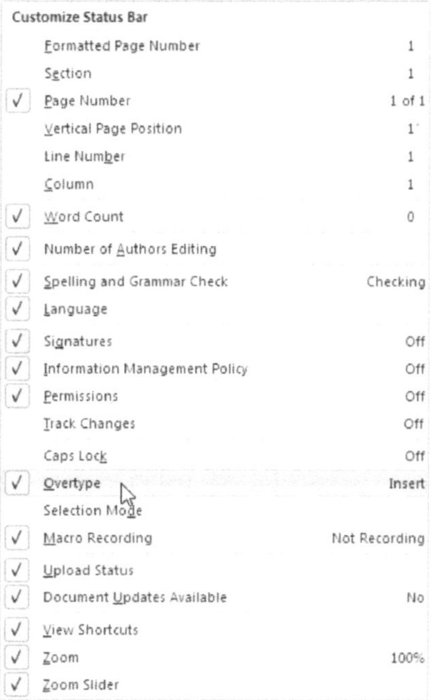

By default, the only way to tell if overtype mode is active is to take a look at the effect your typing has on the information already in your document. If you like to use overtype mode, then you would benefit by informing Word that you want the status bar to indicate the currently active mode. You do this by right-clicking a blank area in the status bar to display the common customizations available. Click the Overtype option, so there is a check mark next to it. With this option turned on, the status bar will display "Insert" if you are currently in insert mode or "Overtype" if you are using overtype mode.

There are three ways to turn on overtype mode. The first is to use the **Ins** key, but this depends on the function you have assigned to that key. The other method is to follow these steps:

1. Display the File tab of the ribbon and then click Options. Word displays the Word Options dialog box.
2. Click Advanced at the left side of the dialog box.
3. Click on the Use Overtype Mode check box, in the Editing Options group.
4. Click on OK.

The final way to turn on overtype mode is to click on the Insert/Overtype indicator on the status bar. Obviously, this method will work only if you've turned on the indicator, as described earlier in this section.

Removing Blank Pages at the End of Your Document

When you print your document, you may notice that a blank page or two prints at the end. If this is the case with your printout, chances are good that you have a bunch of empty paragraphs at the end of your document. Even though you can't see them on the screen (normally), they are there and take up extra space.

To delete the empty paragraphs, simply press **Ctrl+End** to go to the end of your document. Then press the **Backspace** key until you delete all the non-printing characters (such as empty paragraphs). You can tell when you are done by the position of the insertion point. If it is located next to the last character you want printed in your document, you are done deleting.

If you want to see the empty paragraphs that are at the end of your document—before you delete them—you can instruct Word to display non-printing characters. Exactly how you do this is described fully in Chapter 1, "Getting Acquainted with Word 2010."

Repeating Your Typing

If you type a word, phrase, or other passage of text, you can use the **F4** shortcut key to repeat the typing at another location in your document. This shortcut key must be used before you use any other Word command or type any other text. This tip is especially helpful when you want to repeat the insertion of a special symbol or if you want to repeat some other action, such as search for a word or phrase.

Copying and Moving Text

It is a rare document in which text doesn't end up getting moved around quite a bit during the writing process. (That, or it is a very short document.) Word provides several different ways you can both move and copy text. These methods are detailed in the following sections.

Moving Text By Using the Mouse

Word incorporates many different tools you can use to edit your document. Perhaps one of the least used but most unique methods of editing your document uses the mouse entirely. You can use your mouse to move text in the same way as you would move another element, such as a graphic. To move text with your mouse, you need to first ensure that the feature is enabled. To do this, follow these steps:

1. Display the File tab of the ribbon and then click Options. Word displays the Word Options dialog box.
2. At the left side of the dialog box click Advanced.

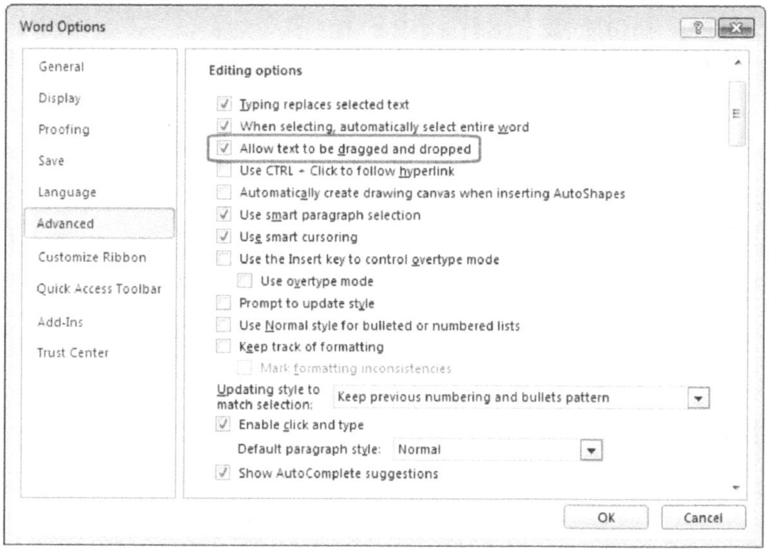

3. Make sure the check box beside the Allow Text to Be Dragged and Dropped option is selected. (The option is in the Editing Options section.)
4. Click OK to close the dialog box.

After the feature is enabled, you can use the mouse to move text by following these steps:

1. Select the text you want to move.
2. Position the mouse pointer over the selected text, and click the mouse button. Make sure to hold it down. Soon, you will notice that some dotted lines appear near the mouse pointer.
3. Drag the highlighted selection to where you want it moved.

4. Release the mouse button. The highlighted text is moved to the location you specified.

If you want to simply copy your text (rather than move it) you can hold down the **Ctrl** key while using the mouse to drag the selection.

There is another way you can move text, and it doesn't rely totally on the mouse. For some people, this method might be a bit easier:

1. Highlight the text you want to move.
2. Hold down the **Ctrl** key and right-click where you want the highlighted text moved.

Don't confuse this method of moving text with the previous information about copying text. Although both use the **Ctrl** key and the mouse, in one of them you are holding down the **Ctrl** key as you drag text, and in the other you hold down the **Ctrl** key as you click at a target location for your text. Although this difference may seem to be subtle, the editing effect on your document is quite different.

Using the Copy Text or Move Text Keys

Many people use the Clipboard to copy text. This technique deletes what was previously in the Clipboard, replacing it with the selection you are copying. At times, however, you may not want to disturb the contents of the Clipboard. Word allows you to bypass the Clipboard and copy text using the keyboard in this manner:

1. Highlight the text you want to copy.
2. Press **Shift+F2**. The message *Copy to where?* appears on the status bar.
3. Move the insertion point to where you want the copy.
4. Press **Enter**.

If you want to move text instead copying it, all you need to do is use **F2** instead of **Shift+F2**.

Moving a Section Break

As you work with Word, you will undoubtedly have a need to move a section break at one time or another. Section breaks in Word are treated the same as any other character. As described earlier in this chapter (see the section "Removing a Break") section breaks appear as a thin double line from one margin of your document to the other.

To move a section break, follow these steps:

1. Select the section break just as you would select any other text.
2. Press **Ctrl+X**. The section break is cut from your document.
3. Position the insertion point where you want to insert the section break.
4. Press **Ctrl+V**. The section break is inserted in your document.

This approach to moving section breaks works with any other type of break as well.

Copying Headers and Footers

If you have developed two documents that are closely related (perhaps they are even different versions of the same information), you may want to copy headers or footers from one document to the other. This is easy to do using standard editing techniques:

1. Select the first document.
2. Display the Insert tab of the ribbon.
3. Click either the Header or Footer tool in the Header & Footer group. Word displays a list of options.
4. Click either Edit Header or Edit Footer, depending on which you want to do. Word switches to Print Layout view (if necessary) and positions the insertion point within the header or footer. The design tab of the ribbon is also displayed.
5. Use the controls in the Navigation group to display the header or footer you want to copy.
6. Select all the elements (text and graphics) in the header or footer.

7. Press **Ctrl+C**. This copies the header or footer information to the Clipboard.
8. Select the second document.
9. Display the Insert tab of the ribbon.
10. Click either the Header or Footer tool in the Header & Footer group. Word displays a list of options.
11. Click either Edit Header or Edit Footer, depending on which you want to do. Word switches to Print Layout view (if necessary) and positions the insertion point within the header or footer. The design tab of the ribbon is also displayed.
12. Use the controls in the Navigation group to display the header or footer where you want to paste your header or footer that you copied in step 7.
13. Position the insertion point in the header or footer, then press **Ctrl+V**. The information in the Clipboard will be inserted in the header or footer.
14. Close the header or footer area for each of your documents. (Click the Close button at the right of the Design tab of the ribbon.)

Working with Headers and Footers

Word allows you to add headers and footers to your document. ***Headers*** and ***footers*** are nothing more than repeating text that appears at the top (headers) or bottom (footers) of every page of your printed document. The beauty of headers and footers is that they have to be defined only once, and then Word places them on your printed pages automatically.

In the following sections, you learn how to use both headers and footers expertly. You start with creating them and then move to positioning them, making changes, and modifying on which pages they appear. You finish up these sections by learning how to insert the current date in your header or footer.

Creating Page Headers and Footers

Headers and footers can add a professional touch to your documents. They are great for "framing" your document and giving each of your pages a cohesive look. The easiest way to create a header or footer in Word is to simply double-click at the top or bottom of a page, where the header or footer will appear. You can also, if you prefer, follow these steps:

1. Display the Insert tab of the ribbon.
2. Click the Header tool in the Header & Footer group, and then click Edit Header. The header is displayed and the insertion point is within it. Word also displays the Design tab of the ribbon.
3. If desired, enter the information for the header, formatting it as you would format normal text in your document.
4. Click on the Go To Footer tool on the Design tab of the ribbon.
5. If desired, enter the information for the footer, formatting it as you would format normal text in your document.
6. Click on Close Header and Footer (on the Design tab of the ribbon) when your header and footer are defined the way you want them.

Positioning Headers and Footers

Now that you know how to create headers and footers, you should know that Word allows you to control where the header or footer appears on each page of your document. To position the header or footer, follow these steps:

1. Position the insertion point in the section whose headers or footers you want to position.
2. Make sure the Page Layout tab of the ribbon is displayed.
3. Click the small icon at the bottom-right of the Page Setup group. Word displays the Page Setup dialog box.
4. Make sure the Layout tab is selected.
5. In the Header box enter a measurement indicating how far from the top edge of the paper you want the bottom line of your header to be printed.

6. In the Footer box enter a measurement indicating how far from the bottom edge of the paper you want the top line of your footer to be printed.
7. Use the Apply To drop-down list to select the part of your document to which your adjustments should apply.
8. Click on OK.

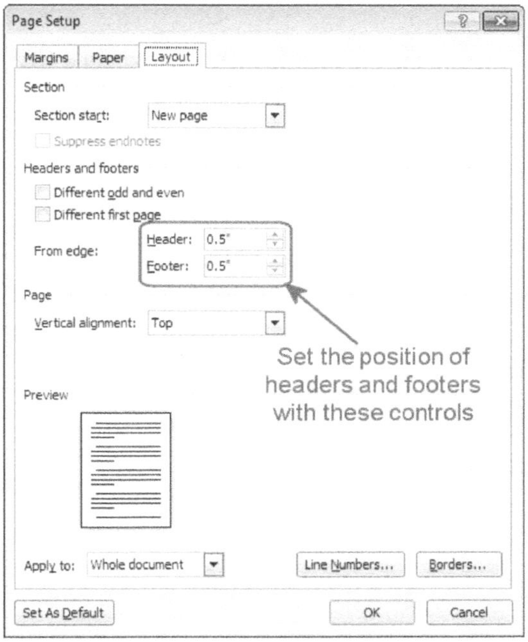

You should take care when positioning headers and footers—if the values you use are too large, you might intrude within the text area of your page. To prevent this, make sure that the values are smaller than the margin settings defined for the page.

Changing Headers and Footers

You can easily change the text in a header or footer by using the tools on the ribbon, in the following manner:

1. Display the Insert tab of the ribbon.
2. Click either the Header or Footer tool in the Header & Footer group. Word displays a list of options.
3. Click Edit Header or Edit Footer, depending on what you selected in step 2. Word activates either the header or footer area of the page.
4. Change the contents of the header or footer in the same manner as you edit normal text in your document.
5. Click on the Close Header and Footer tool (on the Design tab of the ribbon) when you are done making changes.

Note that if you are using Print Layout view, your headers and footers already appear on your screen. To modify the header or footer, you can simply double-click on the area you want to change.

Working with Odd and Even Headers and Footers

Word doesn't limit you to a single header and footer per document. You can also specify that different headers and footers are used for both odd and even pages. Word then automatically determines which footer is the proper one to use for the given page. To define separate odd and even headers or footers, do the following:

1. Display the Page Layout tab of the ribbon.
2. Click the small icon at the bottom-right of the Page Setup group. Word displays the Page Setup dialog box.
3. Click on the Layout tab.
4. In the Headers and Footers section, make sure the check box labeled Different Odd and Even is selected.
5. Click on OK.

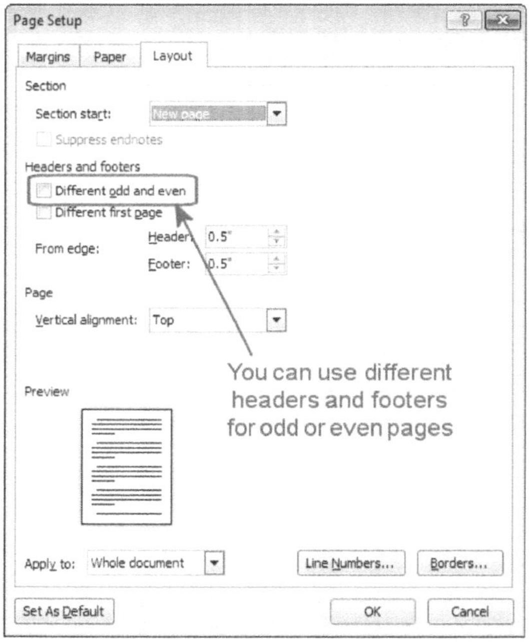

Now you can set the headers and footers, using techniques you learned earlier in this chapter. You can also, if desired, specify different odd- and even-page headers or footers on the Design tab of the ribbon. (Remember: This tab is visible only when you are already working within a header or footer.)

You can also use the Previous and Next tools to move to different headers and footers you've defined for your document.

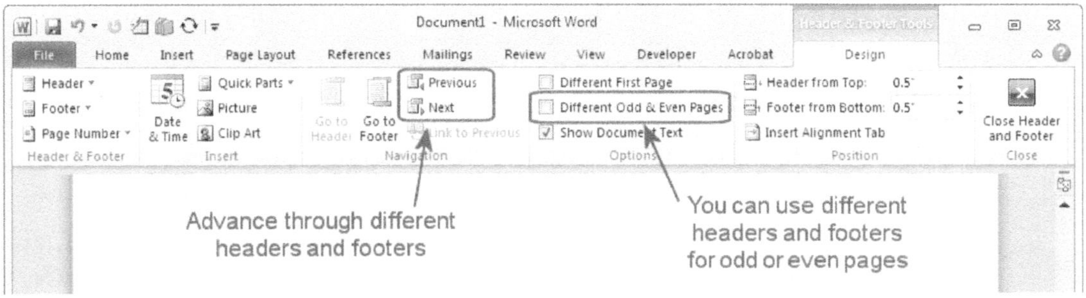

Note that if you have only one page in your document, you can set the header or footer for only the single odd page you have created. If you want to set it for even pages, you need to have at least two pages in the document.

Suppressing Headers or Footers

Many people who come to Word from a WordPerfect environment are curious about how to suppress headers and footers in documents. Although WordPerfect makes this task very easy, the method used in Word is much more flexible.

Word allows you to create headers and footers for each section of your document. Thus, if you have eight sections in your document, each section can have its own header and footer. By default, headers and footers for any new section are the same as in the section just before it. You can change that, however, so that no header or footer appears in a particular section. Thus, it appears that the header or footer is "suppressed," when in fact you have just deleted it for a particular section of your document.

If you want to suppress the header or footer in a particular section of your document, follow these steps:

1. Position the insertion point within the section whose header or footer you want to suppress.
2. Double-click on either the header or the footer, depending on which one you want to suppress. Word dims the main portion of the document, places the insertion point in the header or footer, and displays the Design tab of the ribbon.
3. Make sure the Link to Previous tool, in the Navigation group of the ribbon, is not selected. (If it is not selected, it looks grayed out.)

4. If you are working in the last section of your document, skip to Step 8.
5. Click the Next Section tool in the Navigation group to display the next section's header or footer.
6. Again, make sure the Link to Previous tool is not selected.
7. Click the Previous Section tool (also in the Navigation group) to again display the header or footer for the section for which you want headers or footers suppressed.
8. Delete the header or footer.
9. Click the Close Header and Footer tool at the right side of the ribbon.

Now, if you preview your document by using Print Preview, you can see that headers or footers are in each section except the one where you deleted them.

Deleting Headers or Footers

If you want to delete the headers or footers, follow these steps:

1. Position the insertion point in the section that has the header you want to delete.
2. Display the Insert tab of the ribbon.
3. In the Header & Footer group, click either Header or Footer, depending on which one you want to delete. Word displays a drop-down list of options.
4. Click Edit Header or Edit Footer, depending on which tool you clicked in step 3. Word displays the header or footer along with the Design tab of the ribbon.
5. Select the contents of the header or footer.
6. Press the **Del** key or press **Ctrl+X**.
7. On the Design tab of the ribbon click on Close Header and Footer.

You may wonder why you don't just click Remove Header or Remove Footer, options available after you complete step 3 of these steps. The reason is simple—in my testing, clicking those options removes the header or footer text, but it doesn't remove the header or footer paragraph. Thus, if you have (for instance) a header or footer that includes a paragraph border, that border is still visible after you use the Remove Header or Remove Footer tools. The only way to get rid of the entire thing is to follow the steps as outlined.

Putting Document Names in Headers or Footers

When you print your document, it is often a good idea to include the document name in the header or footer. Then, at a later date, you can recall the name of the file in which the document is stored. To include document names in a header or footer, follow these steps:

1. Display the Insert tab of the ribbon.
2. Click either the Header or Footer tool in the Header & Footer group. Word displays a list of options.
3. Click Edit Header or Edit Footer, depending on what you selected in step 2. Word activates either the header or footer area of the page.
4. Position the insertion point where you want to include the filename.
5. The Insert tab of the ribbon should still be displayed. Click the Quick Parts tool (in the Text group) and then choose Field. Word displays the Field dialog box.
6. From the Categories list select the Document Information category.
7. From the Field Name list, select FileName.

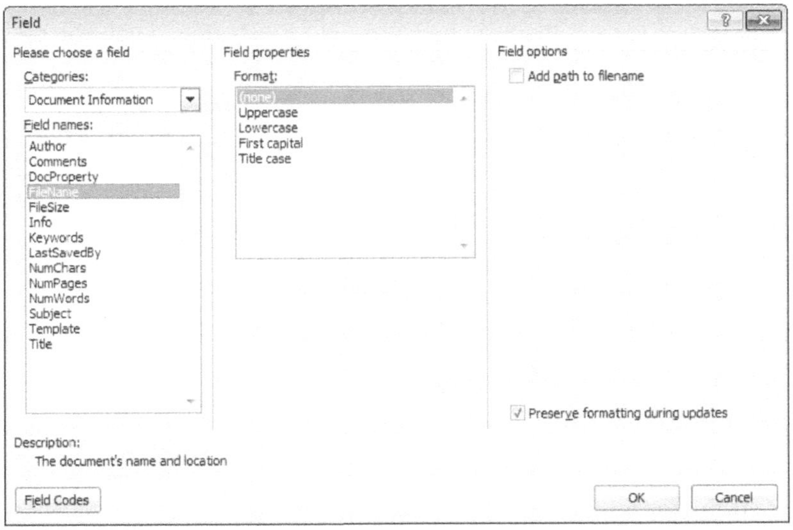

8. Choose the Add Path to Filename option, if desired.
9. Click on OK. The filename of the current document is inserted at the insertion point; it will automatically change if you save the document under another name.
10. Click on Close.

Inserting the Date in Headers or Footers

When people create headers or footers for documents, they commonly insert the date into them. In this way, the date is printed on every page of your document. To insert the date, you can follow these steps:

1. Display the Insert tab of the ribbon.
2. In the Header & Footer group, click either Header or Footer, depending on which one you want to edit. You'll see a list of options from which you can select.
3. Select either the Edit Header or Edit Footer option from those available. Word activates the header or footer area of the page, depending on your choice.
4. Position the insertion point within the header or footer at the place you want the date to appear.
5. Make sure the Design tab of the ribbon is displayed. (It should be displayed by default after completing step 3.)
6. In the Insert group click the Date & Time tool. Word displays the Date and Time dialog box.

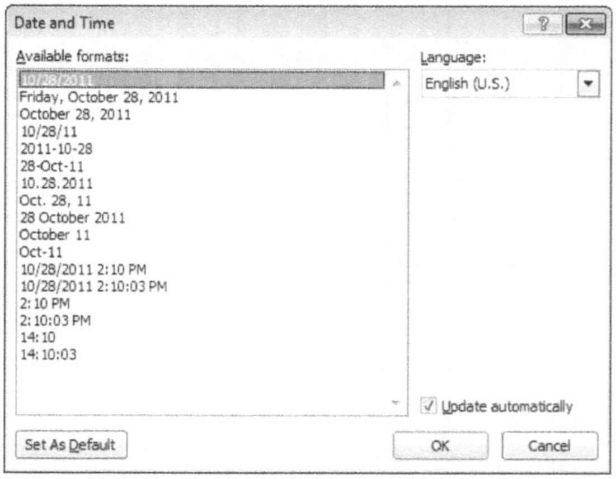

7. Select a date format from those at the left of the dialog box. Word inserts today's date.
8. Click OK to close the Date and Time dialog box.
9. At the right side of the Design tab, click the Close Header and Footer tool.

The date inserted in step 3 is the current date, meaning that the date used is ***dynamic.*** Today, it shows today's date, and tomorrow it will show tomorrow's date. When you use the Insert Date tool, Word inserts a DATE field into the header or footer. You can see the makeup of the DATE field by simply selecting it in the header or footer and pressing **SHIFT+F9**. The field is expanded, and you can make changes to the field.

The Date & Time tool isn't the only way to insert a date in the header or footer, of course. After you position the insertion point where you want it (step 4 in the preceding steps), you can choose the Date and Time tool from the Insert tab of the ribbon and select the format you want used for the date.

Working with Page Numbers

Even though you may use Word to create single-page documents, sometimes you write documents that are quite a bit longer. In those instances, you may want to add page numbers to the document. Page numbers are helpful on printouts because the numbers allow your reader to keep the final printout in the proper order.

In the following sections you learn the two ways you can add page numbers to your documents. You also learn how to modify the starting page number, change the type of characters used for page numbers, and add additional information to page numbers.

Adding Page Numbers

If you create documents that are more than a page or two long, you probably want to include page numbers so that the reader of your printed document can keep it in order. Word supports automatic numbering of pages in your document, and you can control the type of page number, the starting number, and the placement of the page number. In addition, you can include page numbers in the headers and footers on your pages (this topic is discussed in the next section).

The easiest way to add page numbers is to use the Insert tab of the ribbon. In the Header & Footer group you'll notice the Page Number tool. When you click the tool you are presented with a variety of ways in which you can insert your page numbers. These options are presented in several groups:

- Top of Page. The options in this grouping allow you to add a formatted page number to the top of each page (the header) in the current section.
- Bottom of Page. The options in this grouping allow you to add a formatted page number to the bottom of each page (the footer) in the current section.
- Page Margins. Adds a page number to either the left or right margin of the page.
- Current Position. Used to add a page number at the current location of the insertion point.

All told there are well over 100 different ways that you can insert formatted page numbers in your document. All you need to do is pick the one you want, and Word does the rest.

Adding Page Numbers in Headers or Footers

If you already have headers and footers defined for your document, you can add a page number by simply editing the header or footer and then using one of the choices available by choosing the Insert tab of the ribbon, clicking the Page Number tool, and then choosing Current Position. The page number you choose is inserted at the location of your insertion point.

If the standard, run-of-the-mill page number does not fit your discriminating tastes, you may want to add page numbers to your headers or footers yourself. These types of page numbers can be rather fancy, if you like. For instance you may want a page number to say "***Page 37***" (special formatting with the word ***Page*** before the actual page number). To add page numbers in your headers and footers, follow these steps:

1. Position the insertion point in the section of your document that you want to include page numbers.
2. Open the header or footer area of the page you are viewing. You can do this by double-clicking within the header or footer areas. You can also display the Insert tab of the ribbon, click the Header or Footer

tools (as desired), and then choose Edit Header or Edit Footer (as appropriate).

3. In the header or footer, type the word **Page** followed by a space.
4. Click on the Page Numbers tool and then click Current Position. From the resulting options, select the page number format that most closely corresponds to what you want to see. (I normally just choose Plain Number since I'm going to do my own formatting anyway.)
5. Format your header or footer text as you would format normal text in your document.
6. Click on Close.

Changing the Starting Page Number

Word gives you quite a bit of flexibility in working with page numbers. There may be times when you want to change the starting page number for a particular part of your document. For instance, your document may include a couple pages of preliminary information, but you want the page numbering to start on the third physical page. This task is quite easy to do in Word.

Page numbering can be varied by section in your document. Thus, to change the way in which pages are numbered in the preceding scenario, perform the following steps:

1. Position the insertion point at the end of the second physical page in your document.
2. Display the Page Layout tab of the ribbon.
3. In the Page Setup group, click Breaks and then click Next Page. The section break appears in your document, and the text following the break will begin at the top of a new page.
4. Display the Insert tab of the ribbon.
5. With your insertion point located on the third physical page (in other words, right after the section break you added), click Page Number in the Header & Footer group, and then click Format Page Numbers. Word displays the Page Number Format dialog box.

6. In the Page Numbering area of the dialog box (at the bottom), click on the Start At radio button and indicate the starting page number for the section.
7. Click on OK to close the Page Number Format dialog box.

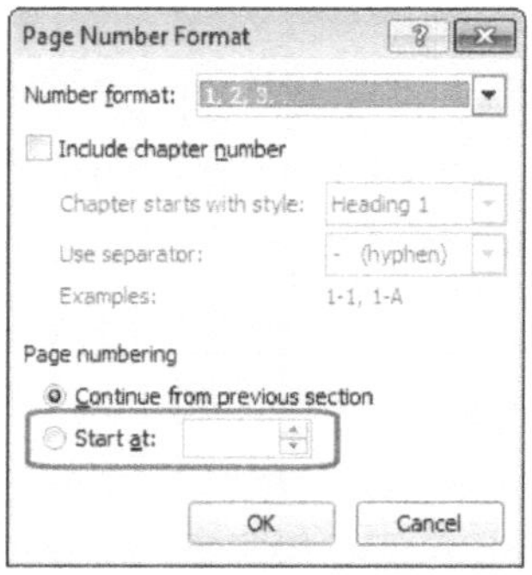

At this point, you can add your page numbers as you normally would. For most people, this means changing the header or footer to include the page number. Remember also that headers and footers can vary by section of your document. Thus, this new section can include the page number in the header or footer, and the preceding section can omit it.

Changing the Type of Page Numbers Used in Headers or Footers

As you already know, Word allows you to make page numbers a part of your headers or footers. After adding the page numbers, you may want to change the type of page numbers used. Word allows you to use several different types of page numbers, as shown in this table:

Type	Selection
Arabic	1, 2, 3, ...
Arabic with dashes	- 1 -, - 2 -, - 3 -, ...
Uppercase roman	I, II, III, ...
Lowercase roman	i, ii, iii, ...
Uppercase alphabetic	A, B, C, ...
Lowercase alphabetic	a, b, c, ...

To change the format of your page numbers, perform the following steps:

1. Position the insertion point in the section of your document whose page numbers you wish to change.
2. Display the Insert tab of the ribbon.

3. Click the Page Number tool, in the Header & Footer group. Word displays some page number options.
4. Choose Format Page Numbers. Word displays the Page Number Format dialog box.
5. Select the numbering format you wish to use from the Number Format list. The table (above) indicates the available types of formats.
6. Click on OK.

Working with Letters and Numbers in Page Numbers

In many documents, you may have a need to number your pages with a section number as a preface. For instance, you may want them numbered as A-1, A-2, and A-3. Exactly how you do this depends on how you have your documents set up. For instance, if each "section" is a different document file, you simply need to set your page numbers this way:

1. Display the Insert tab of the ribbon.
2. Click either the Header or Footer tool in the Header & Footer group, depending on where you want your page numbers to appear. Word displays a list of options.
3. Click Edit Header or Edit Footer, depending on what you selected in step 2. Word activates either the header or footer area of the page.
4. Position the insertion point where you want the page number to appear.
5. Type the section letter followed by a dash. (For instance, **A-** or **B-**.)
6. Make sure the Design tab of the ribbon is displayed.
7. In the Header & Footer group click Page Number | Current Position | Plain Number.
8. Format your header or footer as desired.
9. Click the Close Header and Footer tool.

If all the sections of your manuscript are in a single document file, just make sure you have a real section break between each of your sections. You can then set up your page numbers this way:

1. Display the Insert tab of the ribbon.
2. Click either the Header or Footer tool in the Header & Footer group, depending on where you want your page numbers to appear. Word displays a list of options.
3. Click Edit Header or Edit Footer, depending on what you selected in step 2. Word activates either the header or footer area of the page.
4. Position the insertion point where you want the page number to appear.
5. Again display the Insert tab of the ribbon.
6. Click the Quick Parts tool (in the Text group) and then choose Field. Word displays the Field dialog box.
7. From the Categories list select the Numbering category.
8. From the Field Name list, select Section.

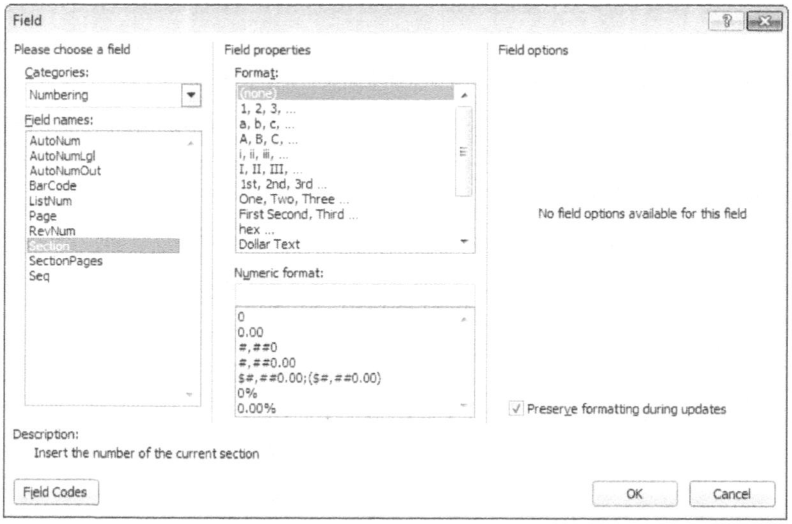

9. Choose the Add Path to Filename option, if desired.
10. Click on OK. The filename of the current document is inserted at the insertion point; it will automatically change if you save the document under another name.
11. In the Format list, choose the option for capital letters (A B C...).
12. Click on OK to actually add your field to the header or footer.

13. Type a dash.
14. Make sure the Design tab of the ribbon is displayed.
15. In the Header & Footer group click Page Number | Current Position | Plain Number.
16. Format your header or footer as desired.
17. Click the Close Header and Footer tool.

4

Searching and Replacing Information

Looking for information in a document can sometimes be like searching for the proverbial needle in a haystack. Fortunately, Word provides a powerful search-and-replace tool that you learn to rely on heavily. In this chapter, you learn the basics of both searching and replacing.

You should know that Word provides two different modes you can use when searching and replacing information. Regular mode—the one that people most often use—is covered in this chapter. The other mode, which uses what Word calls ***wildcards,*** is quite advanced and is not covered in this book. After you master the basics of searching for and replacing text, you may want to refer to back issues of ***WordTips,*** Word's online help system, or a more advanced book on using Word.

Conducting Searches

The searching capabilities of Word allow you to easily find text and other elements within your document, such as formatting, styles, fields, and inserted objects. Even if you want to replace information in a document, you must first learn to search for information. The search tool provided with Word is deceptively simple—which means that you can easily overlook some of the more powerful features it provides. In the following sections, you get a taste for the many different ways in which Word allows you to search for information.

Starting a Simple Search

Word 2010 makes it extremely easy to find information in a document, particularly when what you want to search for isn't that demanding. The best way to start a search is simply to press **CTRL+F**. When you do, Word displays the Navigation task pane at the left side of your document and waits for you to start typing what you want to find.

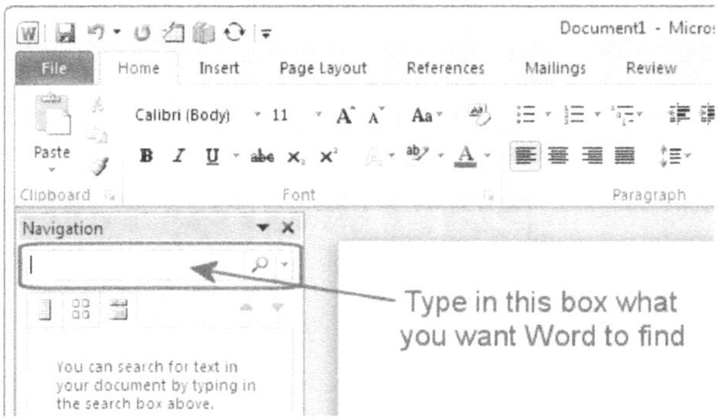

As you type, Word displays whatever matches your text in the Navigation task pane. You can, at any time, click one of the matches and Word takes you to that spot in the document. The really cool thing about this approach is that the Navigation pane provides some context to what it finds, meaning that it shows some of the text just before and just after the text for which you are searching.

Starting an Advanced Search

Sometimes what you need to find cannot be located with just a simple search. In that case, you need to perform what Word 2010 refers to as an advanced search. There are two ways you can instigate such a search:

- Click the down-arrow to the right of the search entry field in the Navigation task pane and then click Advanced Find.
- Display the Home tab of the ribbon, click the down-arrow at the right of the Find tool (in the Editing group) and then choose Advanced Find.

Either way, Word displays the Find tab of the Find and Replace dialog box. The Find tab of the Find and Replace dialog box includes several different controls, but not all

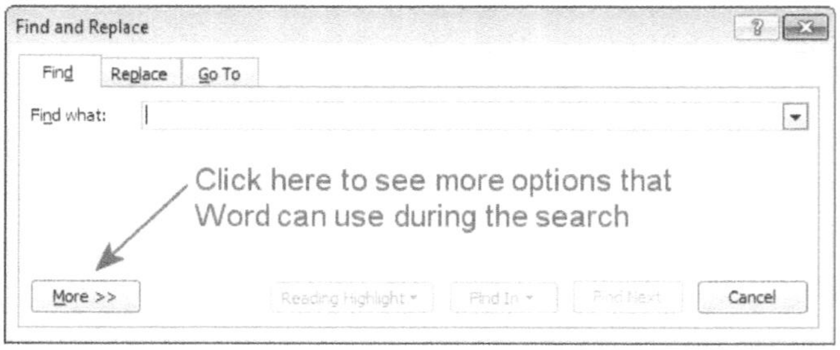

of them may be visible at first. To see all the controls, make sure you click on the More button, if it is available.

Initiating a New Search

Sometimes, you can really muck things up when searching for information. This problem usually happens when you have been searching for text and formatting and then you want to search for something new. If you don't remember to reset all the search variables, you may never find what you are searching for. To overcome this problem, make sure you answer the following questions:

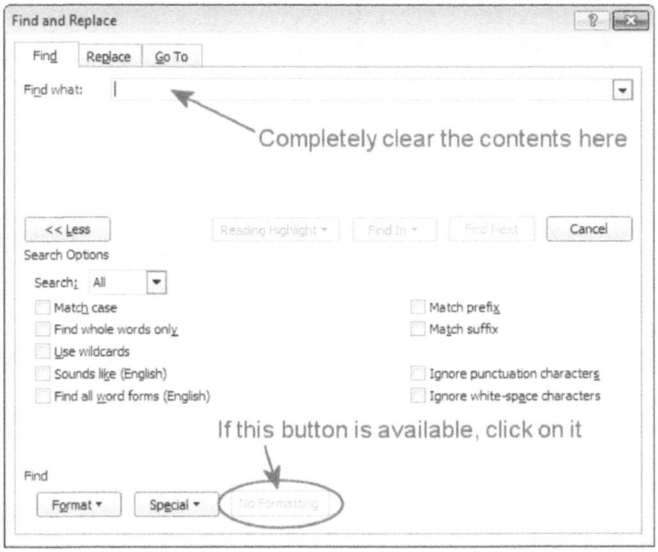

- Did you clear the formatting? If the No Formatting button is available at the bottom of the Find and Replace dialog box, you should click on it. (To see the button, click on the More button, if it is available.)
- Did you clear the text in the Find What box? If not, do so, or at least replace it with text for which you want to search. Remember that a space or some other nonprinting character may be in the box. You should highlight the field and press the **Del** key.

If you still have problems finding what you want, the quickest solution is to exit Word, restart the program, reload the document, and search again.

Searching with a Fast Find-Next

Word provides a fast, efficient, and flexible searching feature. Using Word's searching capabilities, you can look for just about anything in your document.

I find myself using the Find and Replace dialog box quite often in the course of creating a document, but one thing really bothers me: I hate having the Find and Replace dialog box block part of my document as I am stepping through occurrences of a search string. Clicking on Find Next works great, but that bothersome dialog box still blocks my view.

To overcome this problem, I generally do the following when I am searching for something:

1. Use the Find and Replace dialog box as normal, specifying what I want to search for and then looking for the first occurrence of the string.
2. When the first occurrence is displayed, press the **Esc** key or click on Cancel. Word closes the Find and Replace dialog box.
3. To find the next occurrence, press **Shift+F4**.

This procedure works the same as clicking Find Next repeatedly and is just as fast, but it gets rid of the annoying Find and Replace dialog box.

Replacing a Long Block of Text

Word has a very powerful search-and-replace function. Sometimes, you may want to replace short blocks of text with long blocks of text or replace text with a nontext item (such as a graphic). To do this, follow these steps:

1. Copy to the Clipboard whatever you want to use as a replacement. (Easiest way: Select the material and press **Ctrl+C**.)
2. Press **Ctrl+H**. Word displays the Replace tab of the Find and Replace dialog box.

3. In the Find What box, enter the text you want replaced.
4. In the Replace With box, enter **^c**. This step instructs Word to use the Clipboard contents as the replacement.

5. Click on Find Next.
6. Make replacements as desired.

This technique is great if your replacement text contains more than 255 characters, because the contents of the Replace With box must be fewer than 255 characters.

Understanding When Replace Doesn't Work

When you try to replace text in your document using the Replace function, things may not always work. If they don't, check these factors:

- Have you made a selection of text? If so, the replace operation works only within that selection.
- Have you cleared the formatting? (Unless the formatting is necessary for what you are trying to do.)
- Did you spell the Find What and Replace With text correctly?
- Are the search modifiers selected properly? Make sure that Match Case and Whole Word options are set as they should be.

If these factors are set properly and the replace operation still doesn't work properly, you should exit Word and restart the program.

Working with Quicker Multiple-Replace Operations

It is not unusual to use the Replace function multiple times to replace the same information. For instance, you may have a document that contains quite a few tabs. You can use the Replace function to replace two tabs with a single tab. In this way, you can get rid of the extra tabs. However, if your document contains numerous extra tabs, you can replace 8, 20, or 50 tabs with a single tab.

When you are specifying the number of tabs to replace, you should use a multiple of two. For instance, in the Find What box, you enter 4, 8, 16, or 32 tabs; in the Replace With box, you enter a single tab. Continue clicking the Replace All button until you are informed that the requested characters cannot be found.

Next, divide the number of characters in the Find What box by two. Thus, if you had searched for 16 tabs, you would now search for 8. You have to click on the Replace All button only once then. (Word could not find two groups of 8 in a row—if it did, it would have been replaced by the preceding search for 16.) On the next iteration, use only 4 and, finally, 2 tabs. Basically, you have replaced all the extra tabs in four or five operations. This technique is the fastest possible way to do the iterative search-and-replace.

Looking for Special Characters

By this point in the chapter, you probably have a pretty good idea of how to search for simple words and phrases in your document. Word also allows you to add special characters to your search, so you can use the Search tool to find some of the special characters you learned about in earlier chapters of this book.

Searching for ASCII and ANSI Characters

Word allows you to search for any ASCII or ANSI character. (ASCII and ANSI are two different ways to represent characters numerically. Windows—and therefore Word—understands both coding schemes.)

To search for ASCII characters, all you need to do is search for a carat (^) followed by the three numbers representing the ASCII value of the character. For instance, to search for an uppercase A, you could enter **^065**.

Searching for an ANSI character is just as easy, yet subtlety different: You enter a carat (^) followed by a zero (0) and then the three numbers representing the ANSI value of the character.

Searching for Special Characters

Word allows you to search not just for text, but also for special characters that normally do not print. For instance, if you are working with documents that use tabular material, you will often search for tab characters. To search for tabs, all you need to do is search for the characters **^t** (you use a lowercase **t**). Word understands this character sequence as a tab character and will dutifully locate all of them in your document.

Often, you search for other special characters, such as breaks and paragraph marks. Remembering these and typing them can be faster than looking them up all the time. The following is a list of the more common special characters, besides the tab character:

Special Character	Symbol
Paragraph mark	^p
Manual page break	^m
Section break	^b
Column break	^n
Line break	^l
Em dash	^+
En dash	^=
Graphic	^g
Any character	^?
Any digit	^#
Any letter	^$
White space	^w
Caret character	^^

You can also use most of these special characters in the Replace With box when doing a search-and-replace operation. You cannot, however, use the special white space character (^w) in the Replace With box.

Accounting for Formatting

Word doesn't limit you to simply searching for text or special characters. It also allows you to search for any formatting you applied to your text. In the following sections, you learn how to search for both character and paragraph formatting as well as for styles. (You learn more about styles in Chapter 6, "Introducing Styles and Templates.")

Searching for Character Formatting

Word has a very powerful search capability that allows you to find not just text, but also the formatting applied to text. To search for character formatting, follow these steps:

1. Display the Home tab of the ribbon.
2. Click the down-arrow at the right of the Find tool (in the Editing group) and then choose Advanced Find. Word displays the Find tab of the Find and Replace dialog box.
3. Make sure the Find What box is empty.
4. If the More button is available, click it.
5. Click on the No Formatting button, if it is available.
6. Click on the Format button and then select Font. Word displays the Find Font dialog box.

7. Select the character attributes for which you want to search. If a check is in a check box, the text that is matched must have the corresponding attribute. If the check box is empty, the text doesn't have that attribute. A shaded check box means the attribute doesn't matter.
8. Click on OK.
9. Click on Find Next.

Searching for Paragraph Formatting

Word can search for just about any type of formatting you can imagine. In the preceding section, you learned how to search for character formatting. If you want to search for paragraph formatting, follow these steps:

1. Display the Home tab of the ribbon.

2. Click the down-arrow at the right of the Find tool (in the Editing group) and then choose Advanced Find. Word displays the Find tab of the Find and Replace dialog box.
3. Make sure the Find What box is empty.
4. Click on the More button, if it is available.
5. Click on the No Formatting button, if it is available.
6. Click on the Format button and then select Paragraph. The Find Paragraph dialog box appears.

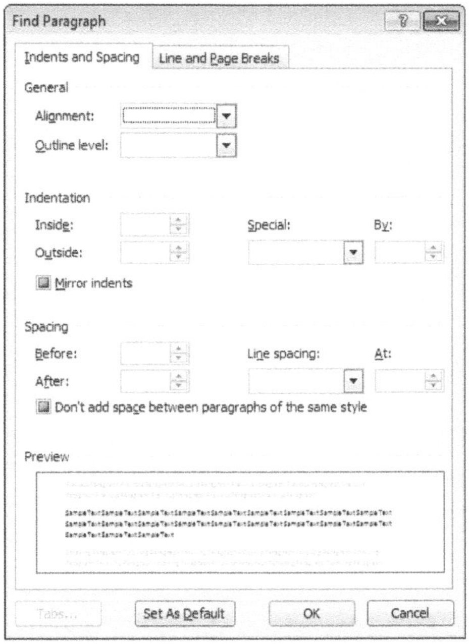

7. Select the paragraph attributes for which you want to search. When you are selecting attributes that use check boxes, a check in a check box means that the matching paragraph must have the attribute. If the check box is empty, the paragraph doesn't have that attribute. A shaded box means that the attribute doesn't matter.
8. Click on OK.
9. Click on Find Next.

Searching for a Style

If you use defined styles in your documents, Word has a very powerful search capability that allows you to find not just text, but also defined styles. To search for a defined style, follow these steps:

1. Display the Home tab of the ribbon.
2. Click the down-arrow at the right of the Find tool (in the Editing group) and then choose Advanced Find. Word displays the Find tab of the Find and Replace dialog box.
3. Make sure the Find What box is empty.
4. Click on the More button, if it is available.

5. Click on the No Formatting button, if it is enabled.
6. Click on the Format button and then select Style. The Find Style dialog box appears.
7. Select the paragraph style for which you want to search.
8. Click on OK.
9. Click on Find Next.

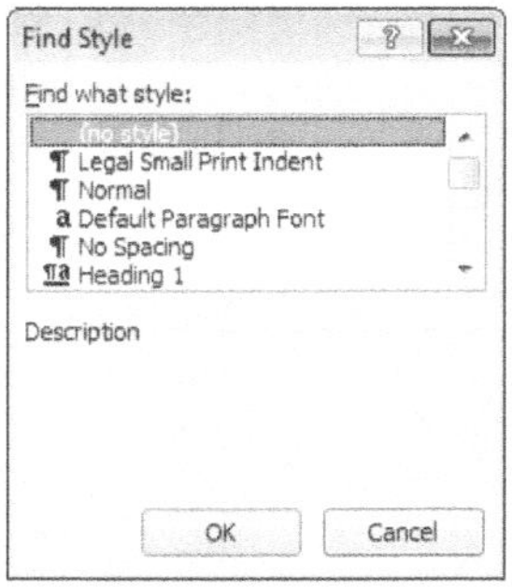

Finding Handy Uses for Search-and-Replace

I hope that now you understand that Word's search-and-replace feature is quite powerful. This power means that you can perform many tasks with search-and-replace to make wholesale changes in your documents.

When putting this book together, I wanted to find a couple of examples of how to use search-and-replace in a constructive way. I came up with a couple that I feel exemplify the types of useful ways you can use search-and-replace and included them in this section. (If you want to see other examples, make sure to consult the *WordTips* web site.)

Updating to Smart Quotes

Smart quotes are those fancy quotes that point in different directions—you see them all the time in typeset material. Word uses smart quotes automatically as one of the features in AutoFormat. However, you may be working with files created in a text editor, a different word processor, or a previous version of Word that did not use smart quotes. In these instances, you may think that you are forced to change the quotes and apostrophes individually (because they are also "smart").

Not so! You can, if you want, use the AutoFormat capabilities of Word to convert the quotes and apostrophes. However, AutoFormat does more than just change quotes and apostrophes, so it can mess up your other formatting. If you want to affect only the quotes and apostrophes and nothing else, follow these general steps:

1. Display the File tab of the ribbon and then click Options. Word displays the Word Options dialog box.
2. Make sure that Proofing is selected at the left side of the dialog box.
3. Click the AutoCorrect Options button. Word displays the AutoCorrect dialog box.
4. Make sure the AutoFormat As You Type tab is displayed.
5. Make sure that the "Straight Quotes" with "Smart Quotes" check box is selected.

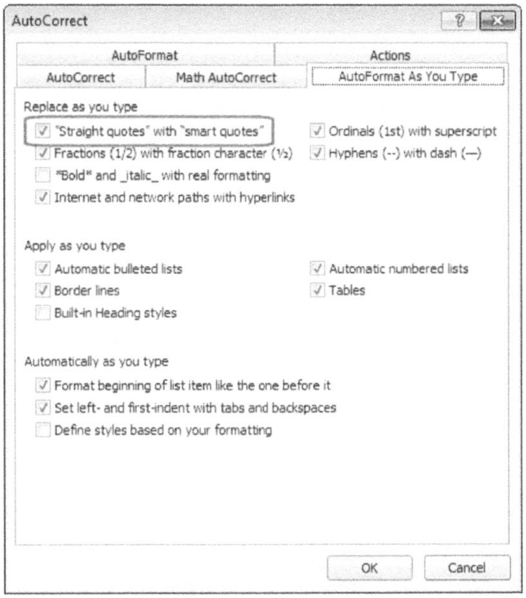

6. Click OK.
7. Press **CTRL+H**. Word displays the Replace tab of the Find and Replace dialog box.
8. In the Find What box, enter a quote mark.
9. Click on the No Formatting button, if it is available. (If the No Formatting button is not visible, you may need to click on the More button to display it.)
10. In the Replace With box, enter another quote mark.
11. Again click on the No Formatting button, if it is available.
12. Click on Replace All.
13. Repeat steps 8 through 12, this time using an apostrophe instead of a quote mark.
14. Close the Replace dialog box when the replacements are finished.

Getting Rid of Smart Quotes

In the preceding section, you learned how to change all the quotes in your document to smart quotes. The antithesis of that task is to convert all quotes to regular quote marks. Sometimes, you may need to send a document to a colleague who doesn't have the smart quotes capability. In these instances, you can get rid of smart quotes and apostrophes by following the same nine steps in the preceding section. The only difference is that in step 5 you need to make sure the Smart Quotes feature is turned off; just clear the check box.

5

Formatting Your Document

After your words are in place just the way you like them, you need to concern yourself with their appearance. This chapter focuses on the primary ways you can use Word to format your text.

Many people try to format their documents as they are writing. This technique is not generally a good idea because your concern with the appearance of your words can distract you from the meaning of your words. You should generally create and edit first and then format later.

Word handles the more common formatting tasks for you—even as you type! For some people, this capability can be quite disconcerting, until they catch on to what Word is doing. In this chapter, you learn not only how to master the automatic formatting features, but also how to set up pages and sections, paragraphs, and individual characters. You also get a crash course in using tabs effectively, using multiple columns in your page layout, turning on line numbers, and applying lists.

Understanding Formatting

Word allows you to format sections (pages), paragraphs, and characters independently of each other. Before you jump into each of these areas, however, you will find a couple of general formatting tips helpful in your journey with Word. In the following sections you learn how to control the Smart Quotes feature of Word and how to change the default font used for your documents.

Automatically Using Smart Quotes

At the end of the last chapter you learned how you can use search and replace to locate and modify smart quotes in your documents. What exactly are smart quotes, however? ***Smart quotes*** is the Word term for what typographers call opening quotes and closing quotes. If you look at the quote marks in any printed book, you will notice that the quote marks at the beginning of a quote look different from those at the end; these are smart quotes.

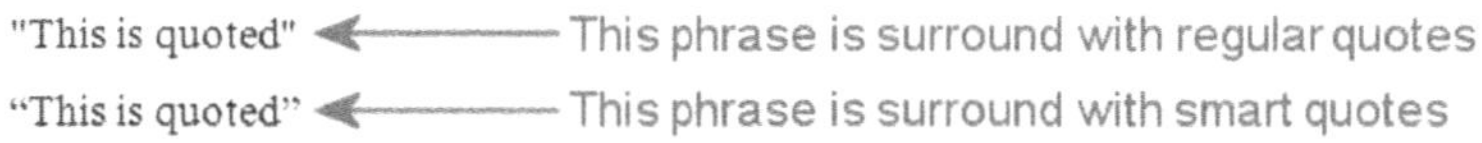

The opening quote is produced by holding down the **Alt** key and typing **0147** on the numeric keypad. The closing quote is produced in the same fashion, except that you type **0148** on the keypad. Because this process can get rather tedious, Word includes an AutoFormat option that changes normal quotes to smart quotes. This option is set by default, but sometimes you may want to check the setting or turn off the option. You can find the option by following these steps:

1. Display the File tab of the ribbon and then click Options. Word displays the Word Options dialog box.
2. Make sure that Proofing is selected at the left side of the dialog box.
3. Click the AutoCorrect Options button. Word displays the AutoCorrect dialog box.
4. Make sure the AutoFormat As You Type tab is displayed.
5. Make sure that the “Straight Quotes” with “Smart Quotes” check box is selected.
6. Click OK.

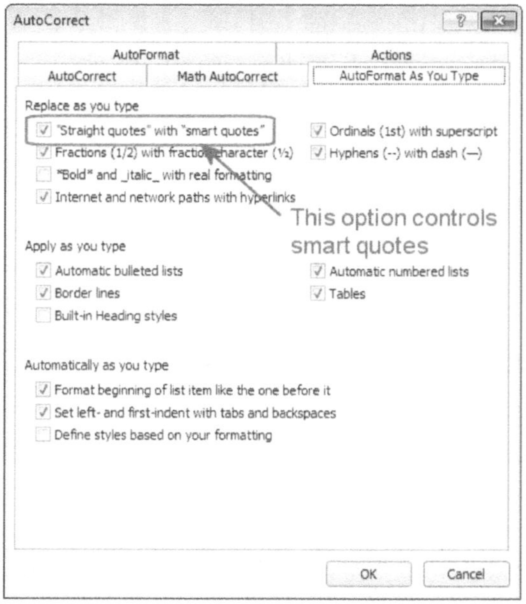

After making sure that this configuration setting is done, just type as you normally would. When you type a quote mark or an apostrophe, Word automatically changes it to its "fancy" state, providing a more professional look to your text.

Changing the Default Font

Every time you create a new document, Word attaches a ***template*** to your file. This template defines, among other things, the default font that is used in your document. (You discover more about templates in Chapter 6, "Introducing Styles and Templates.") If you find yourself changing this font often, you should consider changing the default font. To do this, follow these steps:

1. Create a brand-new blank document.
2. Press **CTRL+D**. Word displays the Font dialog box.
3. Select the font settings you want to use by default.
4. Click on the Set As Default button. Word asks whether you are sure you want to change the default font.
5. Click on Yes.

You have now changed the default font. The changes take effect with the next document you create.

Setting Up Pages and Sections

Those who have used Word for any length of time quickly discover that you can use three general types of formatting for a document: page formatting, paragraph formatting, and character formatting. In general, page formatting is set only once for an entire document. That is because the paper characteristics (such as paper size and orientation) seldom change in the middle of a document.

The following sections describe how you can format your document at a page level. Here, you learn how to determine page layout, change the paper size and orientation, change the margins, and work with sections.

Determining Page Layout

One of the first tasks you must do when planning your document is determining how you want it to look when you are done. This concept is called ***page layout.*** The process of determining page layout is called ***page design.*** To create an effective layout, you must first understand the relationship between the page and the page margins.

First, the ***page*** is the sheet of paper to which you will be printing. ***Margins*** determine how much white space you want to leave around the edges of the paper. Margins can vary within a document, but paper size never varies. You can determine the printable area on a page by adding the margin measurements and subtracting them from the page size. How you determine your layout from there depends on whether you are printing single sheets or facing pages. In printing single sheets, it is assumed that every page is alike. In printing facing pages, you are designing both left and right pages. In this case, margins are typically adjusted to allow for binding of the finished product.

Assume that you will print single sheets. If you are using a page size of 8.5 □ 11 inches and the left and right margins are both 1.25 inches, the horizontal printable area is 6 inches (8.5 minus 2.5, which is the total of the left and right margins). If the top margin is 1 inch and the bottom margin is 1.5 inches, the vertical printing area is 8.5 inches (11 minus 2.5, which is the total of the top and bottom margins).

When you are setting your page margins, you may also need to worry about the ***gutter margin,*** particularly if you are printing facing pages. The ***gutter margin*** is an additional margin added to some page layouts to allow for binding. Thus, in the preceding example, if you were printing facing pages and the gutter margin was .75 inches, the horizontal printing area would be only 5.25 inches (8.25 minus 3.25, which is the sum of the left, right, and gutter margins).

Making Changes to Your Page Layout

When you are working with your page layout, the Page Layout tab of the ribbon is invaluable. It allows you to set many of the characteristics of your page using a rather simple interface.

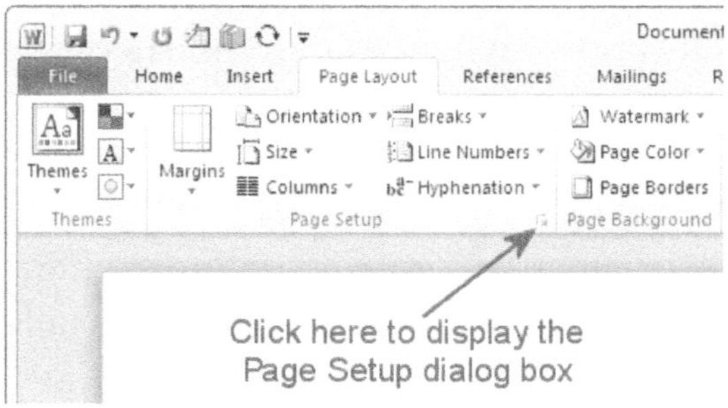

In reality, only the left side of the Page Layout tab directly affects page characteristics. The first three groups (Themes, Page Setup, and Page Background) affect the layout of your page directly. Word can do a lot more relative to page layout than what you see in the Page Layout tab, however. To get the most flexibility in designing your pages, you'll want to learn how to display the Page Setup dialog box; all you need to do is click on the small icon at the lower-right of the Page Setup group and Word displays this powerful dialog box.

Selecting a Paper Size

Word supports any printer supported by Windows. The reason is that Word uses the features offered by Windows to print documents. Many features associated with formatting documents and printing them are related to the type of printer you have installed and selected. For instance, many printers support different sizes of paper.

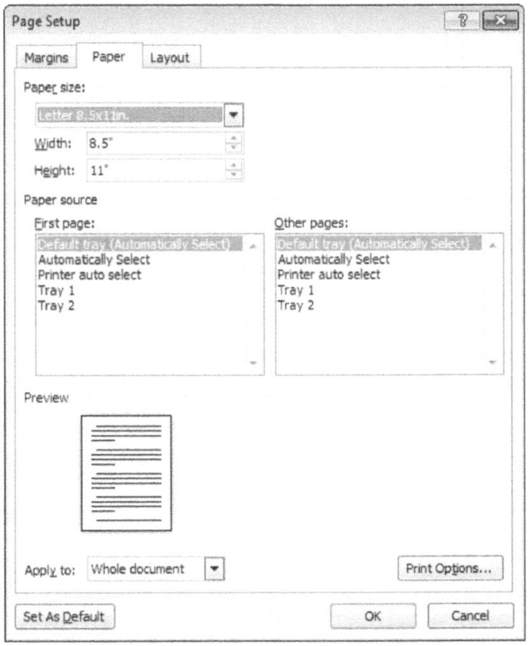

If your printer supports multiple paper sizes and you want to change the paper size used for your document, you can easily do so from the Page Layout tab of the ribbon. Just click the Size tool (in the Page Setup group) and you can choose from the more common paper sizes available. If you need even more options, you can choose More Paper Sizes to display the Paper tab of the Page Setup dialog box.

At this point you can use the controls near the top of the dialog box to specify the exact size of your paper. Remember, though, that the options shown in the dialog box depend on the printer you have installed on your system. Click the OK button when you are finished selecting a paper size.

Changing the Page Orientation

You can use two orientations when you print your document: portrait and landscape. Portrait orientation is taller than it is wide, and landscape is the opposite—wider than it is tall.

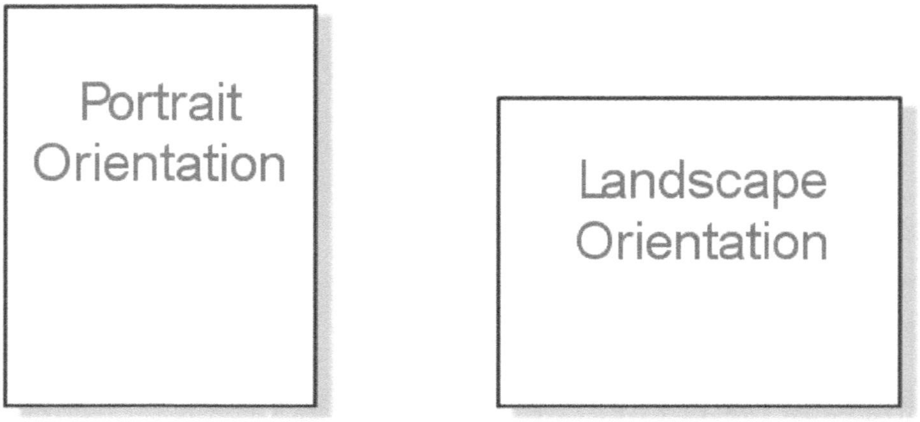

Whether you can modify the paper orientation depends on the type of printer you have installed to work with Windows. You can change the orientation of your page very easily using the Page Layout tab of the ribbon. All you need to do is click on the Orientation tool (in the Page Setup group) and choose which orientation you want to use.

Changing the Page Margins

When you first start Word, certain assumptions are made about your page layout. One assumption has to do with the page margins. You can change your page margins at any time by following these steps:

When you first start Word, certain assumptions are made about your page layout. One of these assumptions has to do with your page margins. You can change your page margins at any time by following these steps:

1. Make sure the Page Layout tab of the ribbon is displayed.
2. Click the Margins tool in the Page Setup group. Word displays a number of common margin settings from which you can choose. (If you see the margins you want, click the setting and you are done.)
3. Click Custom Margins. Word displays the Margins tab of the Page Setup dialog box.
4. Adjust the top, bottom, left, and right margins as desired.
5. Make sure the Apply To drop-down list is set to Whole Document.
6. Click on OK.

Understanding the Gutter Margin

The *gutter margin* is a typographical term used to designate an additional margin added to a page layout to compensate for the part of the paper made unusable by the binding process. In a facing-pages layout, the gutter margin is typically on the inside of both pages.

A gutter margin is not commonly used in a single-sheet layout, although Word allows you to do that. In single-sheet layouts, the gutter margin is typically the area where you might use a three-hole punch on your paper. Word allows you to designate the gutter margin as being on either the left or top side of the paper if you are designing a single-sheet layout. To change the gutter margin, follow these steps:

1. Display the Page Layout tab of the ribbon.
2. Click the Margins tool in the Page Setup group. Word displays a list of potential margin settings.
3. Click Custom Margins. Word displays the Margins tab of the Page Setup dialog box.
4. Using the Gutter control, indicate how wide you want the gutter margin to be.

5. Use the Gutter Position control, if it's available (depending on the Multiple Pages setting), to indicate where the gutter margin should be applied.
6. Make sure the Apply To drop-down list is set to Whole Document.
7. Click OK.

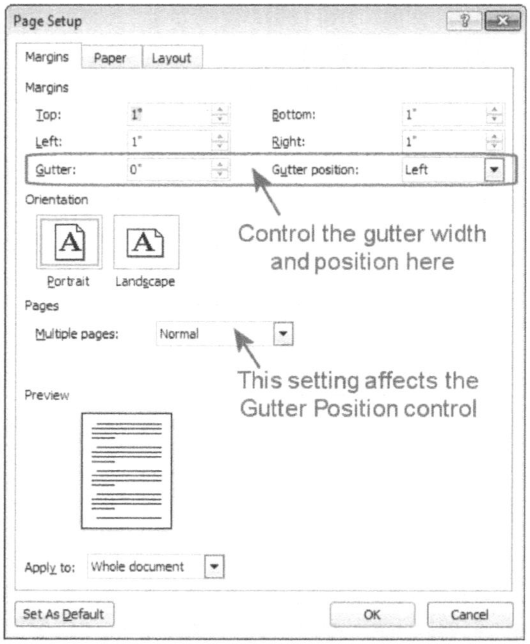

Understanding Sections

Page formatting is typically set once for an entire document. In some cases, you may want to modify page settings in the middle of a document. For instance, you may want to change the top margin on a particular page, or you may want to change the way that headers or footers appear on a particular page.

The way Word handles such mid-document page-formatting changes is through the use of sections. A ***section*** is a portion of a document to which a certain set of page-formatting properties should be applied. If you find that you need to change anything having to do with page layout, simply create a new section and change the formatting for that section alone.

To insert a new section in your document, display the Page Layout tab of the ribbon and use the Breaks tool in the Page Setup group. Word displays two sections in the resulting drop-down list: page breaks and section breaks. It is the section breaks that you need to use in order to change the layout of your pages in the middle of a document. You can select any of the following section breaks:

- **Next Page.** This type of section break causes Word to skip immediately to the next page and begin the new section.
- **Continuous.** This type of section break causes Word to continue with the current section formatting until it makes sense to switch to the

new section formatting. If the new section formatting can be applied immediately, it is. Otherwise, the formatting takes effect on the next page.

- **Even Page.** When this type of section break is entered, Word immediately skips to the next even-numbered page, inserting a blank odd-numbered page, if necessary.
- **Odd Page.** This type of section break causes Word to immediately skip to the next odd-numbered page, inserting a blank even-numbered page, if necessary.

Select the type of section break you want from the drop-down list. The section break is inserted and you can format the new (or old) section as you desire.

Displaying Your Section

If you often work with documents that require multiple sections, you'll find it helpful to instruct Word to display a section number on the status bar. All you need to do is right-click a blank area of the status bar to see a list of different things you can include in the status bar. You want to make sure that a check mark appears next to the Section option, near the top of the list.

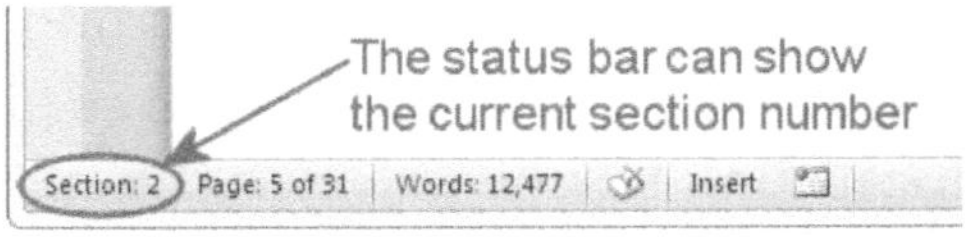

With the indicator displayed you can always tell which section of your document you are in. Just to the left of the page number is the notation ***Section: 1***. It indicates that you are in section 1 of your document. If the notation is ***Section: 3,*** you are in section 3. (You get the idea—you can have as many sections as you need in your document.)

Changing Sections

When you first start using sections, they might seem a bit confusing. For instance, you may have a document that has multiple sections, some formatted as continuous breaks and others as new page breaks. If you later delete some section breaks, it may seem that the section formatting is getting messed up, particularly if you delete the last section break you find in a document.

Why this is happening is best explained by an example. Let's say you have a document that has three sections:

- Section 1 is formatted as New Page.
- Section 2 is formatted as Continuous.
- Section 3 is formatted as New Page.

You must remember that the formatting for each section is maintained in the section break at the end of each section. This concept might seem odd because you can see only two section breaks in the document, even though it has three sections. In the sample document, the following statements are true:

- The formatting for Section 1 (New Page) is in the first section break.
- The formatting for Section 2 (Continuous) is in the second section break.
- The formatting for Section 3 (New Page) is in the implied section break that is always at the end of a document.

Note that an implied section break is at the end of the document. This section break is not visible, of course, but it is nonetheless there, and it contains formatting for the final section.

When you position your insertion point just before the second section break and press the **Delete** key, the section break is deleted. This action deletes the formatting for the second section and the text in that section automatically is formatted according to what was in the third section. Thus, you end up with two sections:

- Old Section 1 (New Page)
- New Section 2 (old Section 3, still New Page)

You may think that Word changed the Continuous section to a New Page section, but in reality Word did exactly what you told it to do: It deleted the second section. You really have no way to delete the final section (the implied section that is always there) in a document. Instead, you must change the formatting of that section so that it reflects what you want.

Avoiding a Section-Break Booby Trap

Word has a "booby trap" that you may not be aware of when you're using section breaks (unless, of course, you have read the preceding section). If you make a section break to create special page formatting and you want (for whatever reason) to remove the section break afterward, you could mess up the formatting of your document.

Let's say that you add a section break to your document and format the portion before the section break differently from the portion after the break. Thus, your document formatting can be described as shown here:

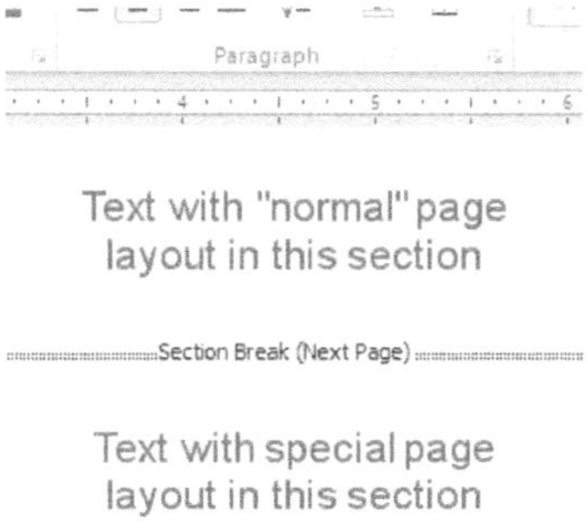

When you delete the section break, the whole document inherits the latter page layout. This result is seldom what you want because you probably want to get rid of the special page layout, not the normal page layout.

One way around this potential problem is to always add a pair of section breaks and then edit the page layout of the middle section:

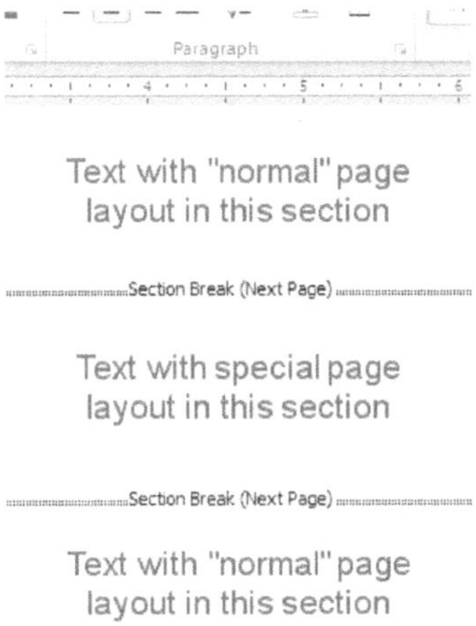

In this case, when you remove both section breaks, you end up with the last page layout (which is "normal").

Formatting Paragraphs

Paragraphs are the elemental building blocks used to put together documents. Word allows you to easily modify the formatting applied to paragraphs. For instance, you can easily set paragraph indents, spacing, and positioning. The following sections describe different ways you can modify the formatting of paragraphs in your document.

Understanding How Word Stores Paragraph Formatting

Word formats paragraphs in two ways: using "hard" (or explicit) formatting or using styles. As a general rule, styles are applied to every paragraph in a document by default. These styles act as a starting point for paragraph formatting. When you make changes to the formatting of a single paragraph, these changes are always made in addition to the formatting inherent in the style used by the paragraph. (Styles are discussed more fully in Chapter 6, "Introducing Styles and Templates.")

When you make formatting changes, the specifications are stored in the paragraph marker at the end of the paragraph. If you have these markers turned on, they appear as a backward P symbol, called a ***pilcrow***. They symbolize a ***hard return***, which is the definition of the end of a paragraph. You can view these paragraph markers by clicking on the Show/Hide tool on the Home tab of the ribbon.

Understanding Paragraph Alignment

One fundamental typographic specification for laying out paragraphs is determining how they appear in relation to the left and right margins. Word refers to this specification as ***alignment***. You can set four types of paragraph alignment within Word:

- **Left-aligned.** All lines in the paragraph butt up against the left text margin. No extra spaces are added to the line. The text of each line does not line up with the right margin, so traditional typesetting terminology often refers to left-aligned text as ragged right.
- **Center-aligned.** All lines in a paragraph are centered between the left and right text margins. No extra spaces are added to the line. The text lines up with neither the left nor right margins.

- **Right-aligned.** All lines in a paragraph butt up against the right text margin. No extra spaces are added to the line. The text of each line does not line up with the left margin, so traditional typesetting terminology often refers to right-aligned text as ragged left.
- **Justified.** All lines in a paragraph are expanded so that they butt up against both the left and right text margins. Space is added as necessary between words and characters to fill out the line. In some typesetting references, justified text is also referred to as full justified.

You can change the alignment of any paragraph by using the appropriate tools in the Paragraph group on the Home tab of the ribbon or by using the Alignment drop-down list in the Paragraph dialog box.

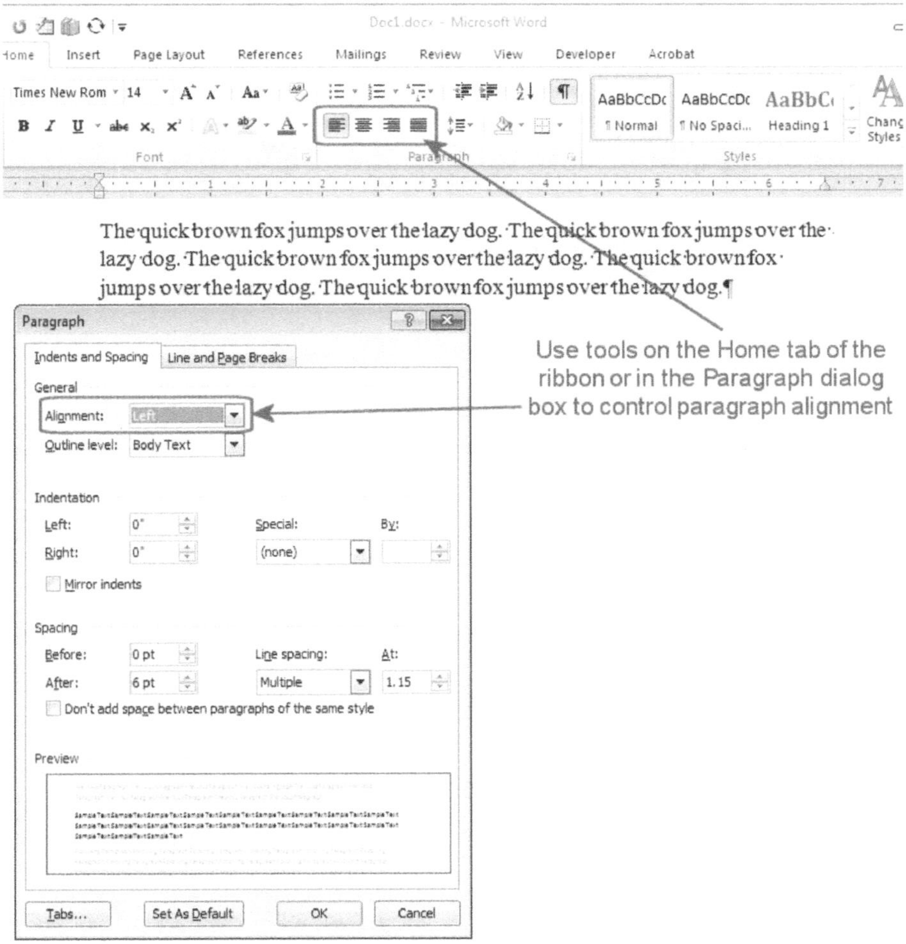

Understanding Line Spacing

Line spacing is the term used by Word to designate the vertical space from the same point on one line to the same point on the next, within the same paragraph. Typically, this distance is measured from baseline to baseline, where the baseline is the nonprinting reference line on which your characters rest.

For those with a typographical background, line spacing is typically measured in points and is referred to as ***leading***. (I tend to use both ***line spacing*** and ***leading*** as synonymous terms.) Word allows you to specify line spacing in points, but more often it is specified in lines.

The default line spacing used by Word (as specified in the Line Spacing drop-down list of the Paragraph dialog box) is single spacing. With this type of line spacing, Word adjusts line spacing based on the largest font size or element on each individual line.

The problem with single line spacing is that it can leave your paragraphs looking uneven and choppy, particularly if you mix fonts and point sizes within the same paragraph. You can attempt to correct this problem by choosing other line-spacing settings, such as 1.5 Lines or Double. The best way, however, is to always specify the Exactly line-spacing option and then set a point size in the At box.

When specifying specific line spacing for your text, you can follow an old typographical convention which states that line spacing should be 120 percent of the point size of your font. Thus, according to this rule, if the font you are using is 10-point Arial, your line spacing should be 12 points. In typographical terms, this setting is referred to as 10/12 Arial (pronounced "10 on 12 Arial"). Likewise, if you are using 28-point Courier, your leading should be 33.6 points. You can round this number to 34 points, even though Word accepts fractional point sizes in the At box.

Understanding At-Least Line Spacing

Word can use several methods for line spacing, as detailed in the preceding section. Typically, the default line-spacing type (as specified in the Line Spacing drop-down list of the Paragraph dialog box) is single spacing. Line spacing is then adjusted, automatically, based on the largest font size or element on each individual line. If you end up with a line that has nothing but small type on it, the spacing for that line is less than for the other lines in the paragraph.

If you don't want this situation to occur, you can use the At Least line-spacing setting in the Paragraph dialog box. This setting allows you to set a minimum line leading for the paragraph. Word is free to increase the leading beyond what you specify, but it cannot reduce it.

Working with Paragraph Formatting Shortcuts

Many people find it easier to use keyboard commands than to constantly stop and use the mouse. Word accommodates these people by providing several shortcuts to format paragraphs. The following table lists several keyboard shortcuts you can use for your paragraph formatting:

Shortcut	Meaning
CTRL+1	Single-space paragraph
CTRL+5	One-and-a-half-line paragraph
CTRL+2	Double-space paragraph
CTRL+L	Left-align paragraph
CTRL+E	Center paragraph
CTRL+R	Right-align paragraph
CTRL+J	Justify paragraph
CTRL+0	Add or remove one line of space before the paragraph
CTRL+M	Increase indent one tab stop
SHIFT+CTRL+M	Decrease indent one tab stop
CTRL+T	Increase hanging indent by one tab stop
SHIFT+CTRL+T	Decrease hanging indent by one tab stop

Putting Flush Left and Flush Right on the Same Line

This quick-and-dirty tip tells how to have seemingly contradictory alignments on the same line. In Word, this trick is done with tabs. In a nutshell, you follow these steps:

1. Make sure the paragraph is formatted as left-aligned.
2. Display the Tabs dialog box. (Display the Home tab of the ribbon, click the small icon at the lower-right corner of the Paragraph group, then click the Tabs button.)
3. Insert a right-aligned tab near the right edge of the line.
4. Click on Set.
5. Click on OK.
6. Type your text. Press the **Tab** key between the information to be left-aligned and the information you want right-aligned.

This trick works great if the information you are formatting is limited to a single line. As an example, this technique can easily work for a chapter name and page number in a header or footer. (You know—the chapter name appears at the left and the page number at the right.)

If you need to accomplish the same task for multiple lines, you should use a small table with two or three cells. The leftmost cell of the table can be used for the left-aligned information, and the rightmost cell can be used for right-aligned information. The center cell—if you choose to use one—is used for spacing purposes. (Working with tables is described in Chapter 7, "Using Tables.")

Adjusting Spacing after a Paragraph

If you use a typewriter—or even some rudimentary word processors—to add space between paragraphs, you simply press the **Enter** key a few times. Word gives you much greater control over the spacing between paragraphs. To modify the amount of space left between the end of one paragraph and the start of the next, follow these steps:

1. Make sure the insertion point is in the paragraph after which you want to adjust spacing.

2. Display the Page Layout tab of the ribbon.
3. Adjust the setting in the Spacing After control (in the Paragraph group) to reflect how much space you want to leave after the current paragraph.

To adjust the value in the Spacing After control, you can use the up and down arrows to the right of the box or enter a value directly. Values can be entered in any of the standard measurements accepted by Word. (The values are automatically converted to points.)

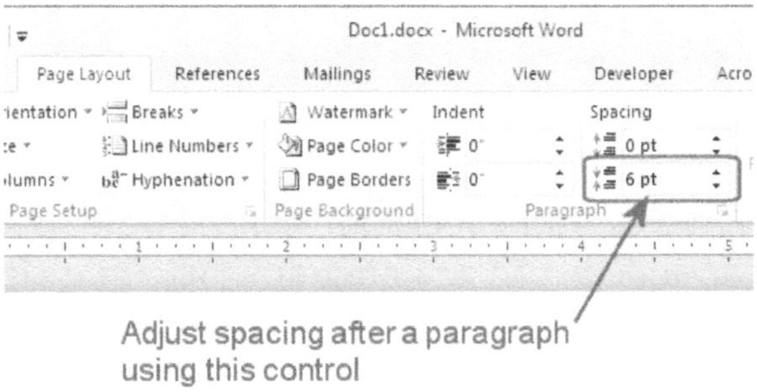

Putting Your Space Before or After

It is not uncommon for page layout designs to call for extra spacing between paragraphs. Word allows you to add extra space either before or after a paragraph. The option you choose is up to you. Remember, however, that the spacing is cumulative. For example, if a paragraph is formatted for one line after it, followed by a paragraph formatted for one line before it, two lines will appear between the paragraphs. You make your design and layout work much easier if you are consistent in where you add the extra space—either before or after a paragraph.

Indenting a Paragraph

If you want to quickly indent a paragraph from the left margin, you can do so by using any of these means:

- Display the Home tab of the ribbon and click on the Increase Indent tool in the Paragraph group. This action moves the left margin to the position of the next tab stop.
- Display the Page Layout tab of the ribbon and use the Indent Left control in the Paragraph group.
- Press **Ctrl+M**. Each time you press it, the left margin is moved to the position of the next tab stop.

- Use the mouse to drag the indent markers on the ruler. Release the mouse button where you want the paragraph indented.
- Use the Paragraph dialog box (Home tab of the ribbon, click the small icon at the lower-right corner of the Paragraph group) and display the Indents and Spacing tab. You can then precisely set the indent in the Left field.

While any of these approaches will work fine, the second one (utilizing the Indent Left control on the Page Layout tab of the ribbon) has the best mix of easiness and precision. You can also use the Indent Right control to adjust the indent of the paragraph from the right margin. You can double-indent a paragraph by adjusting whatever is in both the Indent Left and Indent Right controls.

Decreasing a Paragraph's Indent

One of the common tasks done to "break up" text and make it more interesting to the reader is to indent some paragraphs. For instance, you may have several paragraphs indented to set them off from the left margin of your document. If you want to undo an indented paragraph, or reduce the indent, you can quickly using any of the same methods described in the previous section.

Setting a Negative Paragraph Indent

You probably already know that Word allows you to specify an indent from the left margin on a paragraph-by-paragraph basis, if desired. The indent value you set is added to the left margin value for the page, thereby deriving how far the left side of the paragraph should appear in relation to the left edge of the paper.

For some layout tasks, you may want to set a negative indent for a paragraph. Word allows you to do this, effectively moving the left side of the paragraph into the margin area you defined for your page. To set a negative indent, start by selecting the paragraph or paragraphs you want affected by the formatting change. (If you want the change to only a single paragraph, you can simply make sure the insertion point is within that paragraph.)

Next, use one of the five adjustment techniques described earlier in this chapter (see the section "Indenting a Paragraph"), set a negative indent or drag the indent markers past the margin.

Creating a Hanging Indent

Hanging indent is the typographical term for a paragraph in which the first line is not indented, but subsequent lines in the paragraph are. Typically, hanging indents are used for numbered and bulleted lists. To create a hanging indent in Word, use the following steps:

1. Position the insertion point in the paragraph in which you want the hanging indent.
2. Display the Home tab of the ribbon.
3. Click the small icon at the bottom-right of the Paragraph group. Word displays the Paragraph dialog box.
4. Make sure the Indents and Spacing tab is displayed. (It should be displayed by default.)
5. In the Special drop-down list, choose Hanging. The By field changes to a value, most often one-half inch.
6. Adjust the By field to indicate how much you want each line in the paragraph (except the first) indented.
7. Click on OK.

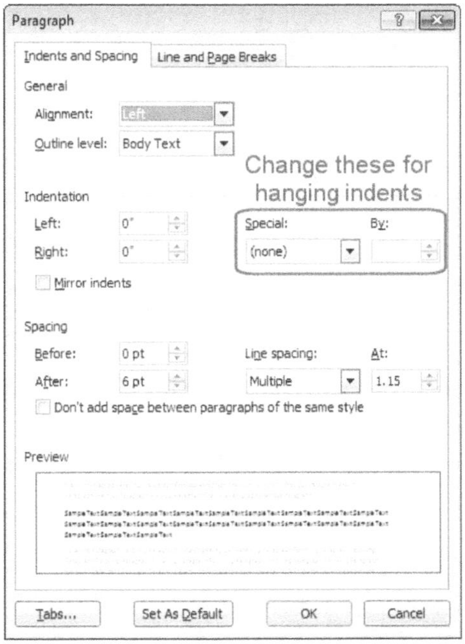

Of course, Word provides shortcuts to create the most common application of hanging indents: numbered and bulleted lists. To apply these, you can simply use the appropriate Numbering or Bullets tools in the Paragraph group on the Home tab of the ribbon.

Forcing a Page Break Before a Paragraph

As you create your documents, you may see certain types of paragraphs that should always begin at the top of a page. For instance, you may use a certain paragraph format for section headings. Your page design also calls for all sections to begin at the top of a page, so you must come up with a way to ensure that this always happens.

You can do this in two ways. The first is to search for each section heading and manually insert a page break before each of the heads. This process can be time consuming, however. Formatting the paragraph so that it always begins on a new page is much easier. Follow these steps:

1. Position the insertion point in the section head paragraph.
2. Display the Home tab of the ribbon.
3. Click the small icon at the bottom-right of the Paragraph group. Word displays the Paragraph dialog box.
4. Click on the Line and Page Breaks tab.
5. Make sure the Page Break Before check box is selected.
6. Click on OK.

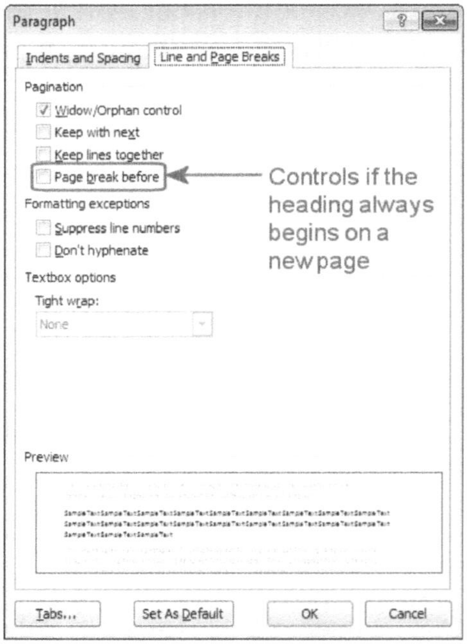

Using page breaks before your section heads is even more automatic if you use a style for all the heads. All you need to do is format the style so that the head always has a page break before it. Then, every time you use that heading style, Word inserts the requisite page break. (Styles are covered in detail in Chapter 6, "Introducing Styles and Templates.")

Preventing Straggling Heads

Certain paragraphs naturally belong with other paragraphs. For instance, section headings belong with the paragraphs that follow them; a signature line belongs with the explanation line that may be after it; a figure belongs with the caption that may follow it. Word makes it easy to keep certain paragraph pairings together. To format paragraphs so that they stay together, use the following steps:

1. Position the insertion point in the first paragraph of the pair you want to keep together.
2. Display the Home tab of the ribbon.
3. Click the small icon at the bottom-right of the Paragraph group. Word displays the Paragraph dialog box.
4. Click on the Line and Page Breaks tab.
5. Make sure the Keep with Next check box is selected.
6. Click on OK.

Keeping Paragraphs on the Same Page

For some types of documents, you may not want your paragraphs to flow smoothly from one page to the next. Instead, you may want to make sure that any given paragraph appears all on one page or another. I require this type of formatting quite often in letters, legal documents, and proposals.

Word includes a formatting feature that allows you to ensure that paragraphs stay together without an automatic page break in the middle of them. You apply this formatting characteristic by following these steps:

1. Select the paragraph or paragraphs you want affected by the formatting change. (If you want the change to only a single paragraph, you can simply make sure the insertion point is within that paragraph.)
2. Display the Home tab of the ribbon.
3. Click the small icon at the bottom-right of the Paragraph group. Word displays the Paragraph dialog box.
4. Make sure the Line and Page Breaks tab is selected.

4. Make sure the Keep Lines Together check box is selected.
5. Click on OK.

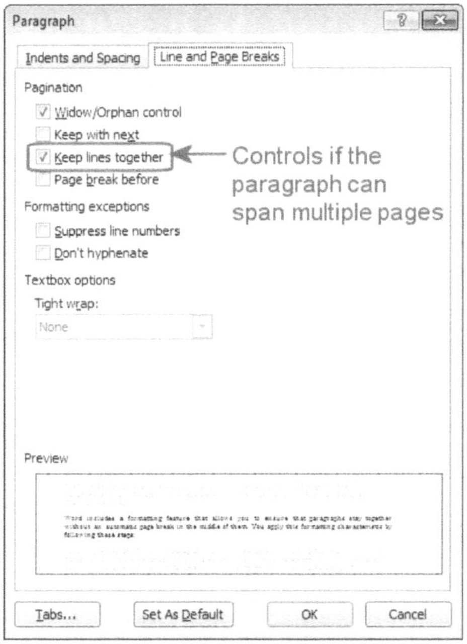

Controlling Widows and Orphans

In typographical terminology, widows and orphans are closely related (no pun intended). These terms refer to one line (and sometimes two lines) of a paragraph left by itself on a page. A ***widow*** is the last line of a paragraph left by itself at the top of a page; an ***orphan*** is the first line of a paragraph left by itself at the bottom of a page. You want to avoid both widows and orphans in your documents because they break up the flow of the text and tend to distract the reader.

Word allows you to automatically control single-line widows and orphans in your documents. To control widows and orphans, follow these steps:

1. Display the Home tab of the ribbon.
2. Click the small icon at the bottom-right of the Paragraph group. Word displays the Paragraph dialog box.
3. Click on the Line and Page Breaks tab.
4. Make sure the Widow/Orphan Control check box is selected.
5. Click on OK.

Resetting the Default Paragraph Formatting

Word provides an easy way to remove any explicit paragraph formatting that was done to a paragraph. This is all you have to do:

1. Select the paragraphs from which you want to remove the formatting.
2. Press **Ctrl+Q**.

The paragraph formatting for the selected text is set back to its default condition. (The default condition is defined by the paragraph formats specified in the style applied to the paragraph.)

Copying Paragraph Formatting

Often, it is beneficial to copy paragraph formatting from one place in your document to another. Word provides a quick way you can do this by using the mouse:

1. Select the paragraph whose formatting you want to copy.
2. Display the Home tab of the ribbon.
3. Click on the Format Painter tool, in the Clipboard group.
4. Using the mouse, select the paragraph you want formatted.

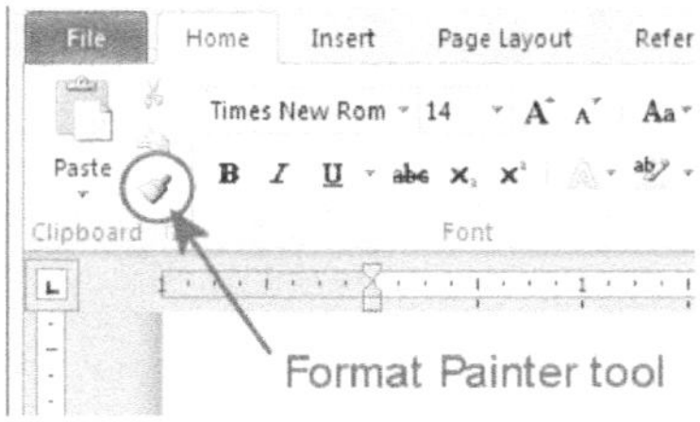

You can also copy the same format repeatedly by double-clicking the Format Painter tool in step 3. You then can apply the same formatting until you turn off format painting by clicking on the tool one more time or by pressing **Esc**.

Adding a Border to a Paragraph

Unlike some word processors, Word allows you to quickly and easily add different types of borders to your paragraphs. You can add borders to any or all sides of a paragraph, or you can create a drop-shadow effect. (***Drop shadows*** cause boxed paragraphs to "stand off" the page, providing almost a three-dimensional appearance.)

One way to add borders to a paragraph is to display the Home tab of the ribbon and then use the Border tool, in the Paragraph group. This tool allows you to easily apply common borders; you can use the down-arrow at the right of the tool to see which types of borders are available.

If you want the most flexibility in applying borders, however, you'll need to rely on the Borders and Shading dialog box. Follow these steps:

1. Position the insertion point in the paragraph to which you want the border added.
2. Display the Home tab of the ribbon.
3. Click the down-arrow next to the Borders tool, in the Paragraph group. Word displays a variety of options you can choose.
4. Choose Borders and Shading. Word displays the Borders and Shading dialog box.

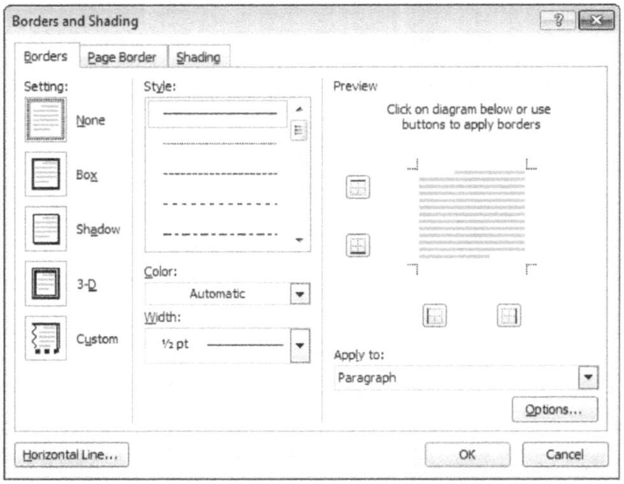

5. Make sure the Borders tab is selected.
6. For a regular boxed appearance around your paragraph, select the Box option from the Setting area.
7. To use a drop-shadow box, select the Shadow option from the Setting area.
8. Word includes a 3-D box that you can use by selecting the 3-D option from the Setting area.
9. If you want a custom border, select the Custom option from the Setting area. Then, you can control the border by clicking on the line buttons located in the Preview box.
10. Select a line type from the Style list.

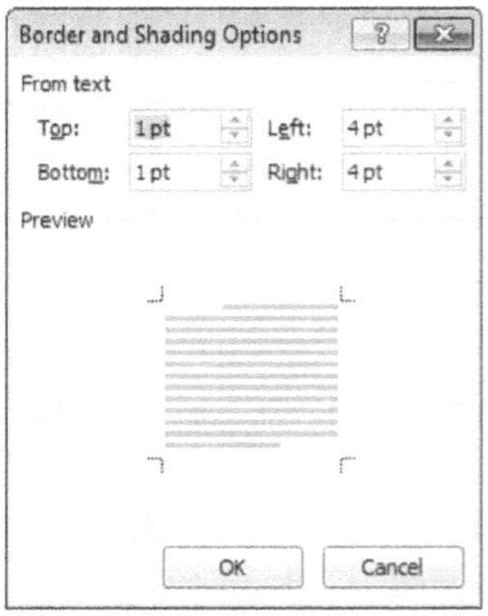

11. To adjust the distance the borders of the box will be from the paragraph text, click on the Options button to display the Border and Shading Options dialog box.
12. Change any of the From Text distances, clicking OK when done.
13. Click on OK to close the Borders and Shading dialog box.

You can quickly remove a paragraph border by selecting None in the Borders and Shading dialog box or No Border from the drop down list displayed in step 2.

Changing Paragraph Borders

In the preceding section, you learned how to add borders to your paragraphs using the presets built into Word. This action results in a paragraph contained in a box in which all four sides match each other. Word allows you to mix and match line types for your border, however. For example, you could have a paragraph that has different types of borders on all four sides. To change borders on your paragraph, follow these steps:

1. Position the insertion point in the paragraph to which you want the border added.
2. Display the Home tab of the ribbon.
3. Click the down-arrow next to the Borders tool, in the Paragraph group. Word displays a variety of options you can choose.
4. Choose Borders and Shading. Word displays the Borders and Shading dialog box.
5. Make sure the Borders tab is selected.
6. Select a line type from the Style list.
7. In the Preview area of the dialog box (the part that shows what your borders look like), click on the side of the paragraph to which you want a border added.
8. Repeat steps 6 and 7 for each of the other paragraph sides.

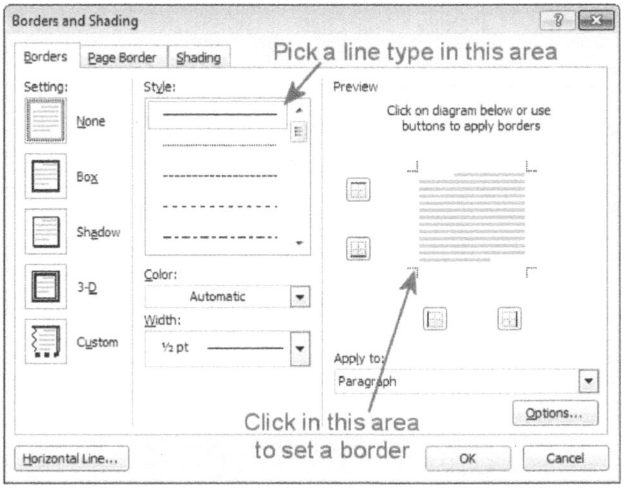

9. Click the Options button, if desired, to adjust how far you want the borders to be from the paragraph text.
10. Click on OK.

Adding a Drop Shadow to a Paragraph

You already know that Word allows you to quickly and easily add borders to your paragraphs. You can take this concept one step further by adding drop shadows to your text. A ***drop shadow*** is a type of paragraph border that is wider on two sides than it is on the other two. The result, especially when printed, is that the bordered paragraph looks like it is raised above the rest of the document.

> This paragraph has a drop shadow applied to it. Notice how the border is different on the bottom and right side.

To add a drop shadow to a paragraph, follow these steps:

1. Position the insertion point in the paragraph to which you want the drop shadow added.
2. Display the Home tab of the ribbon.
3. Click the down-arrow next to the Borders tool, in the Paragraph group. Word displays a variety of options you can choose.
4. Choose Borders and Shading. Word displays the Borders and Shading dialog box.
5. Make sure the Borders tab is selected.
6. Select the Shadow option from the Preset area on the left side of the dialog box.
7. Select a line weight from the Width drop-down list.
8. Click on Options, if desired, to adjust the distance between the borders and the text.
9. Click on OK to close the Paragraph Borders and Shading dialog box.

Deleting a Border

If you add a border around a paragraph in your document, you may want to remove the border later. This is easy to do by following these steps:

1. Position the insertion point in the paragraph from which you want the border removed.
2. Display the Home tab of the ribbon.
3. Click the down-arrow next to the Borders tool, in the Paragraph group. Word displays a variety of options you can choose.
4. Choose No Border from the available options.

Formatting Characters

A document is made of up dozens, hundreds, or thousands of characters. Word allows you to individually format every single character in your document, if you are so inclined. (Gee, you could make your boss's weekly report look like a ransom note!) Word allows you to apply many different types of character formats, such as bold, italics, underline, strikethrough, and many others.

In the following sections, you are introduced to the different ways you can format your text. You learn not only how to apply formatting, but also how to remove it. You also learn what font sizes are and how to use them to your advantage.

Applying Formatting to Words

The general way you accomplish tasks in Word is to select the text on which you want to perform an action and then perform the action. For instance, if you want to make a text selection bold, you first select the text and then click on the Bold tool on the Home tab of the ribbon (in the Font group) or press **Ctrl+B**.

There is one notable exception to this pattern, however: If you want to apply formatting to a single word in your document, you don't need to first select the word. All you need to do is make sure the insertion point is located within the word. To test out this concept, follow these steps:

1. Position the insertion point within a word in your document.
2. Display the Home tab of the ribbon.
3. Click on the Bold tool, in the Font group. The entire word should be changed to bold type.
4. Press **Ctrl+U**. The entire word should become underlined and the Underline tool (near the Bold tool in the Font group) is automatically selected.
5. Press **Ctrl+Space Bar**. The formatting of the word returns to its default setting.

You can try out this technique with virtually any character formatting. It even works if you simply position the insertion point and display the Font dialog box (press **Ctrl+D**). Any changes you make in the dialog box are automatically applied to the entire word. If formatting words in this way does not work on your system, follow these steps:

1. Display the File tab of the ribbon and then click Options. Word displays the Word Options dialog box.
2. Click Advanced at the left side of the screen.
3. In the Editing Options area, make sure the When Selecting, Automatically Select Entire Word check box is cleared.

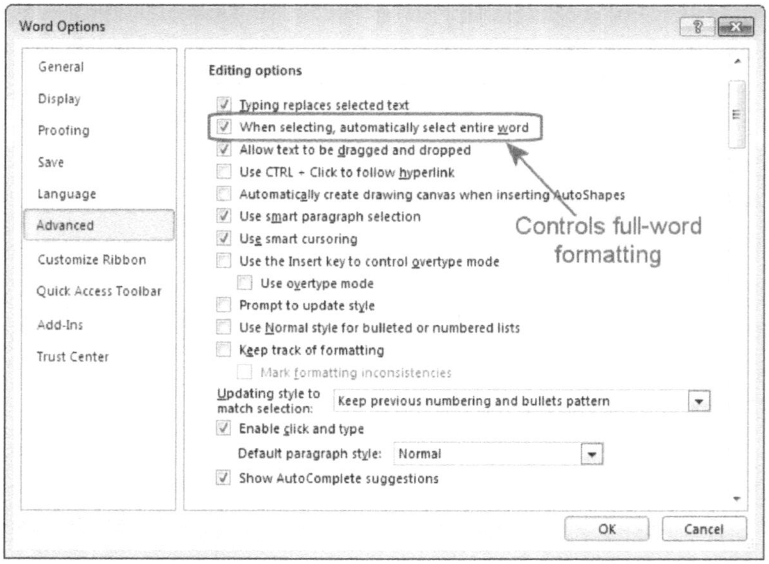

4. Click on OK.

Now the automatic word formatting should operate exactly as just cited.

Quickly Displaying the Font Dialog Box

If you want to quickly format some characters in your document, try this trick:

1. Select the characters you want to format.
2. Right-click on the selected characters. Word displays a Context menu.
3. Select Font from the Context menu. The Font dialog box appears.
4. Format your characters.
5. Click on OK.

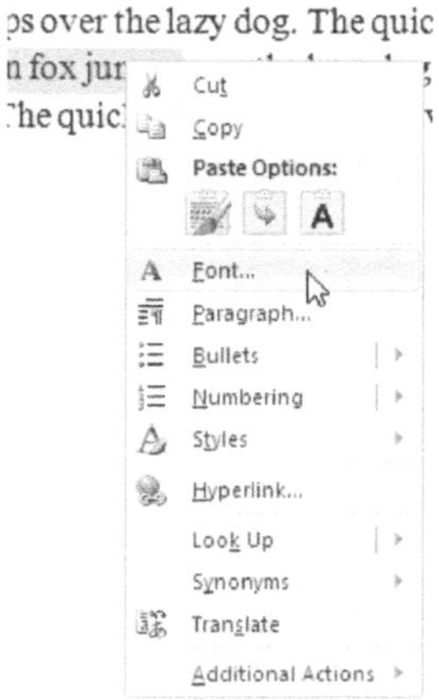

Resetting the Default Character Formatting

Word provides an easy way to remove any explicit character formatting that was done to a group of characters. This is all you have to do:

1. Select the text from which you want the formatting removed.
2. Press **Ctrl+Space Bar**.

The character formatting for the selected characters is set back to its default condition. The characters then take on the default attributes associated with the paragraph's style.

Setting Accurate Font Sizes

You already know that Word allows you to easily change the size of your fonts. You can use either the formatting tools on the various ribbon tabs or the controls in the Font dialog box (press **Ctrl+D**) to set the size of the type you've selected in your document. You may not know, however, that Word can use virtually any point size you want, not just those listed in the drop-down size lists.

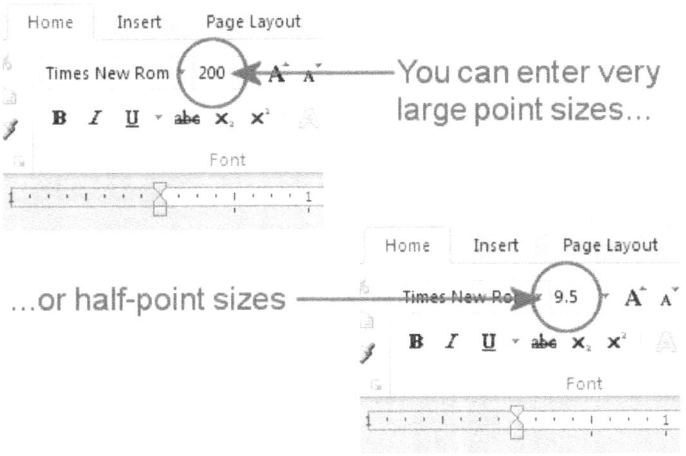

Font sizes are specified in points, which are a typographer's measurement roughly equivalent to 1/72 of an inch. You can either select a size from the drop-down list, or you can enter your own size. Select the size shown in the list, and then type the size you actually want. When you press **Enter**, the size of your selected text is changed. This is great when you want your text to be really large for special signs. For instance, you can easily set your font size to 120 or 200 points to make huge letters for a sign.

It is also easy to overlook the fact that Word can display and print fonts in increments of half a point. Depending on the typeface being used, this can make a big difference. For example, there is a very marked difference between 10 and 11 point Verdana, and 10.5 may be just what you need.

You obtain the half-sizes by typing them directly into the size box. If you try to type in any other fractional size (like 10.25 or 10.4), Word displays an error message. You can only set full point sizes or half point sizes.

Quickly Changing the Point Size of Text

Word provides a way to quickly increase or decrease the point size of a text selection. To do this, follow these steps:

1. Select the text whose point size you want to increase.
2. Press **Ctrl+]** to increase the point size, or press **Ctrl+[** to decrease the point size.

Every time you press **Ctrl+]** (that's the right bracket character), the point size of the selection increases by one point. Every time you press **Ctrl+[** (that's the left bracket character), the point size of the selection decreases by one point.

If you want more precise control over the point size (for instance, to make something 11.5 points), you must use either the controls on the Home tab of the ribbon (in the Font group) or in the Font dialog box.

Quickly Changing the Font Size

Word allows you a great deal of control over the size of the font used in your documents. If you want to change font sizes quickly, you can follow these steps:

1. Select the text whose font size you want to change.
2. Press **Ctrl+>** to increase the size of the font.
3. Press **Ctrl+<** to decrease the size of the font.

Exactly how much the font size is increased or decreased varies. At smaller point sizes (12 or smaller), the point size is changed by a single point. Between 12 and 72 points, you are actually stepping through the point sizes available on the drop-down Font Size tool on the Home tab of the ribbon, in the Font group. (Those sizes are 12, 14, 16, 18, 20, 22, 24, 26, 28, 36, 48, and 72.) Thereafter, the font size is changed by increments of ten points. You can use this method to reduce a point size to a single point or to as large as 1638 points.

Understanding Types of Fonts

Two types of fonts are used in word processing: proportional-space fonts and monospace fonts. Proportional-space fonts are designed so that every letter occupies only the minimum horizontal space necessary for the letter. Thus, an *i* takes less space than a *w*.

Monospace typefaces, on the other hand, are designed so that every letter and character take the same amount of horizontal space. If you have ever spent any time working on typewriters, you are familiar with monospace fonts—all the fonts used by typewriters fall into this category.

Word supports both monospace and proportional typefaces, as long as the font is installed correctly in Windows.

Making Text Bold

You probably already know that you can make text bold in Word in a number of ways. One common way is to press **Ctrl+B** when you want to "turn on" the bold character attribute, type your word (or words), and then press **Ctrl+B** to turn it off.

You can make text bold in another quick way: Use asterisks. This method may sound strange, but it is quite common in the online world. In plain-text e-mail messages, people often use asterisks to *emphasize* words. If you type asterisks in this manner (directly next to a word or group of words, with no intervening spaces) in Word, the asterisks are automatically removed and the word or words are shown in bold type. You can try this technique yourself by typing *this is bold* (exactly as shown here, including spacing) within Word. When you type the second asterisk, both asterisks disappear and the words "this is bold" are shown in bold type.

If this formatting method doesn't work on your system, follow these steps:

1. Display the File tab of the ribbon and then click Options. Word displays the Word Options dialog box.
2. At the left side of the dialog box click Proofing.
3. Click AutoCorrect Options button. Word displays the AutoCorrect dialog box.
4. Make sure the AutoFormat As You Type tab is displayed.
5. Make sure the *Bold* and _ Italic_ with Real Formatting check box is selected.

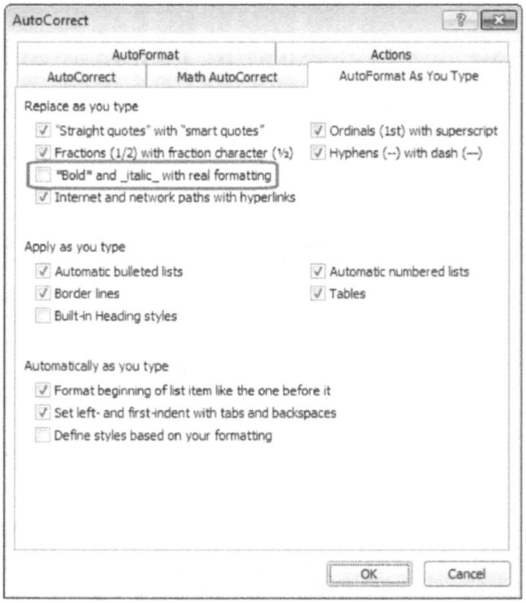

6. Click OK to dismiss the AutoCorrect dialog box.
7. Click OK to dismiss the Word Options dialog box.

The formatting should now work as described.

Understanding Underlines

You can use several different types of underlines in Word. The underlining of characters is handled by pressing **Ctrl+D** to display the Font dialog box. In the Underline Style drop-down list, you can select any of eighteen underline types.

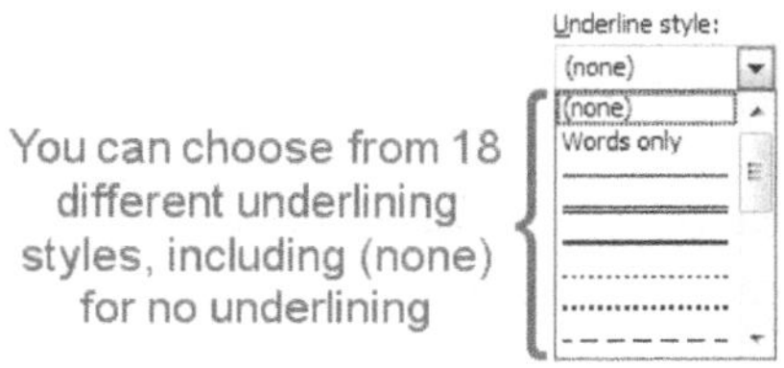

To see all the underline types available you can scroll down through the options presented in the Underline Style drop-down list.

Copying Character Formatting

Often, it is beneficial to copy character formatting from one place in your document to another. Word provides a shortcut for doing this task:

1. Select the characters whose formatting you want to copy.
2. Display the Home tab of the ribbon.
3. Click on the Format Painter tool, in the Clipboard group.
4. Select the text you want formatted.

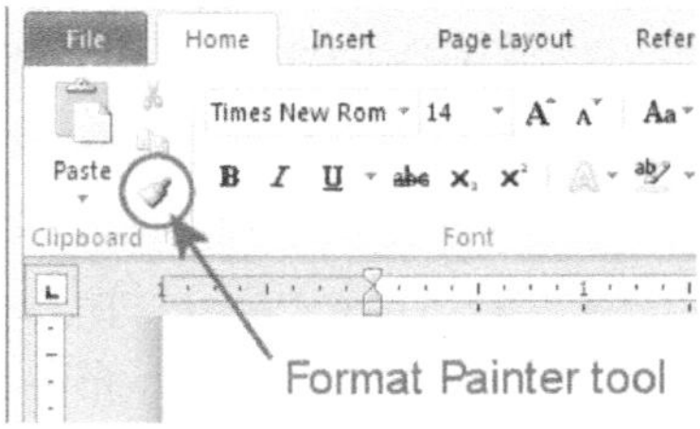

You can also copy the same format repeatedly by double-clicking the Format Painter tool in step 3. You then can apply the same formatting until you turn off format painting by clicking on the tool one more time or pressing **Esc**.

Changing the Text Case

You've probably had it happen to you: You get a document from the new temporary employee down the hall (or the technically illiterate mid-level manager) and you need to get it ready for a presentation in ten minutes. When you open the document, you see that EVERYTHING IN THE DOCUMENT IS SCREAMING AT YOU. All the text is in capital letters. Aaagghhh! You can't distribute the document in this format. Quick—what do you do?

Fortunately, Word allows you to quickly and easily change the capitalization (case) of text. All you need to do is follow these steps:

1. Select the text you want to alter.
2. Press **Shift+F3**. Word changes the case of the selected text.
3. Continue pressing **Shift+F3** until the case is the way you want it.

Using the **Shift+F3** method allows you to cycle through three different case scenarios: ALL CAPS, all lowercase, and All Sentence Case (first letter capitalized). If you need greater control, you need to use the tools on the ribbon. Simply select your text and then display the Home tab of the ribbon. In the Font group, click the Change Case tool. Word displays a list of five different ways you can transmogrify your text:

- **Sentence Case.** This option capitalizes the first letter of each sentence in the selection.
- **Lowercase.** This option makes all the selected text lowercase.
- **Uppercase.** This option capitalizes each letter in the selection.
- **Capitalize Each Word.** This option capitalizes the first letter of each word in the selection.
- **Toggle Case.** This option switches the case of each letter in the selection: Lowercase becomes uppercase and vice versa.

Working with Tabs and Tab Stops

A common way to format information is to use tab stops. Tab stops allow you to quickly and accurately align information on a line in your document. If you are familiar with typewriters, you are already familiar with the concept of tab stops. Word goes far beyond the rudimentary tab stops in typewriters, however. It allows you to set four major types of tab stop:

- **Center.** Text entered after pressing the **Tab** key is centered horizontally at the position where the tab stop is located.
- **Decimal.** Text entered after pressing the **Tab** key is aligned horizontally so that the decimal point (or period) is where the tab stop is located. If there is no decimal point, text is right-aligned to the tab stop.
- **Left.** Text entered after pressing the **Tab** key is aligned horizontally so that the left side of the text is where the tab stop is located.
- **Right.** Text entered after pressing the **Tab** key is aligned horizontally so that the right side of the text is where the tab stop is located.

In the following sections, you learn ways to set, change, clear, and delete tab stops. You also learn some special tips for dealing with default tab stops and how to place a tab stop at the beginning of a paragraph.

Setting Tab Stops By Using the Tabs Dialog Box

Word provides two major ways to set tab stops. One involves using the ruler, which is best deferred to the next section. The other involves using the Tabs dialog box. To set tab stops using this method, use the following steps:

1. Make sure the insertion point is in the paragraph for which you want to set tabs stops.
2. Display the Home tab of the ribbon.
3. Click the small icon in the lower-right corner of the Paragraph group. Word displays the Paragraph dialog box.
4. Click the Tabs button. Word displays the Tabs dialog box.
5. In the Tab Stop Position box, enter the measurement for where you want the tab stop located.
6. Click on the Alignment type desired.
7. Click on the type of Leader desired, if any.
8. Click on Set.

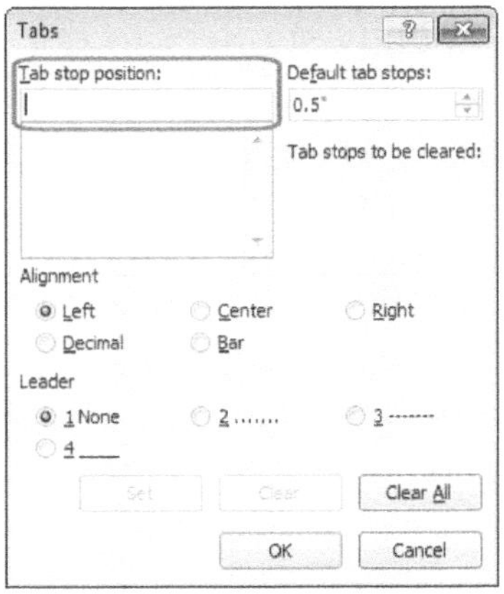

9. Repeat steps 3 through 6 for each tab stop desired.
10. Click on OK.

Notice in step 7 that Word allows you to set leaders for your tab stops. ***Leaders*** are characters that are inserted in the otherwise blank area created by the tab. Leaders are often used with right-aligned tabs or decimal tabs, particularly when the text column is quite wide and the tab covers a wide area. Notice that Word allows you to specify four different types of leaders.

You may also have noticed from examining the Tabs dialog box a tab alignment type known as Bar. This is not really a tab stop, but rather a way to place a vertical bar at a particular location in your paragraph. In fact, referring to Bar as an alignment type is a misnomer because no text is aligned.

Changing Tabs By Using the Ruler

After tabs are set, they appear on the ruler, right under the measurement markings. You can quickly adjust tab stops by using your mouse to drag the markings to a new location on the ruler. Simply point to a tab stop marking with the mouse, click on the left mouse button, and drag the tab stop to a different location. Release the mouse button when you are satisfied with the new tab location.

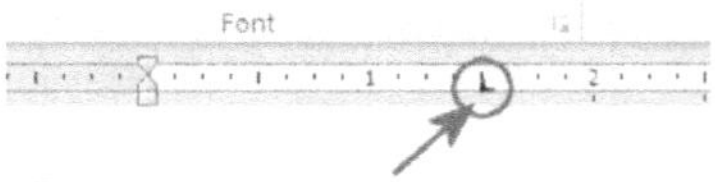

Tab stops are noted on the ruler

Quickly Displaying the Tabs Dialog Box

A handy shortcut to quickly display the Tabs dialog box is to double-click on the ruler just below the markings that appear there. Make sure you double-click at the bottom of the ruler, not at the top. (If you double-click at the top, the Page Setup dialog box appears.)

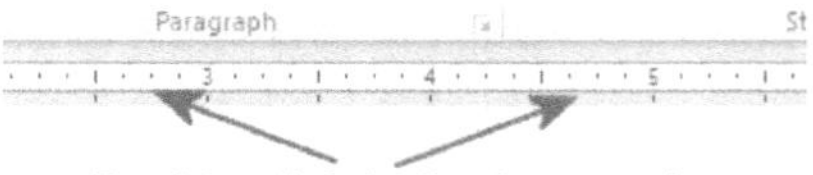

Double-click in the lower ruler area to display the Tabs dialog box

You should note that when you double-click on the ruler, you may also inadvertently set a tab stop. If you want to delete the newly placed tab stop, you can do so at the Tabs dialog box.

Clearing All Tab Stops

Sometimes, you may want to clear all tab stops for a certain paragraph. For instance, you may decide that you want to reset all the tabs—it is just easier to start with a clear slate, so to speak. To clear all the tab stops for a paragraph, follow these steps:

1. Make sure the insertion point is in the paragraph whose tab stops you wish to clear.
2. Display the Home tab of the ribbon.
3. Click the small icon in the lower-right corner of the Paragraph group. Word displays the Paragraph dialog box.
4. Click the Tabs button. Word displays the Tabs dialog box.
5. Click on Clear All.
6. Click on OK.

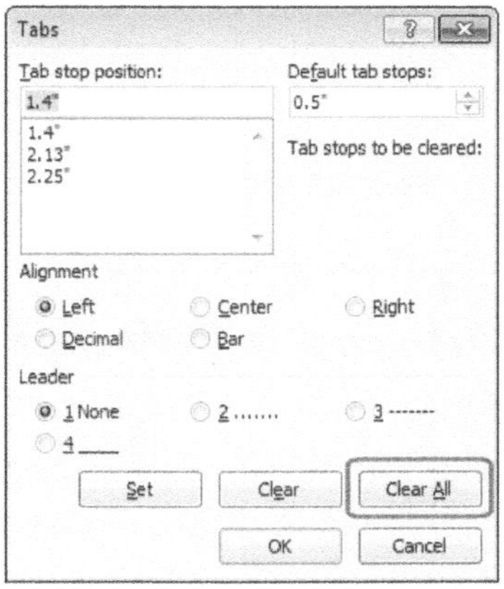

If you want to clear all tab stops in the entire document, simply press **Ctrl+A** in place of step 1. Word selects the entire document.

Deleting a Tab Stop

Sometimes, you may not want to clear all the tab stops—you want to clear only a few of them. To clear individual tab stops for a paragraph, follow these steps:

1. Make sure the insertion point is in the paragraph whose tabs you want to clear.
2. Display the Home tab of the ribbon.
3. Click the small icon in the lower-right corner of the Paragraph group. Word displays the Paragraph dialog box.
4. Click the Tabs button. Word displays the Tabs dialog box.
5. On the tab list below the Tab Stop Position box, select the tab stop you want to clear.

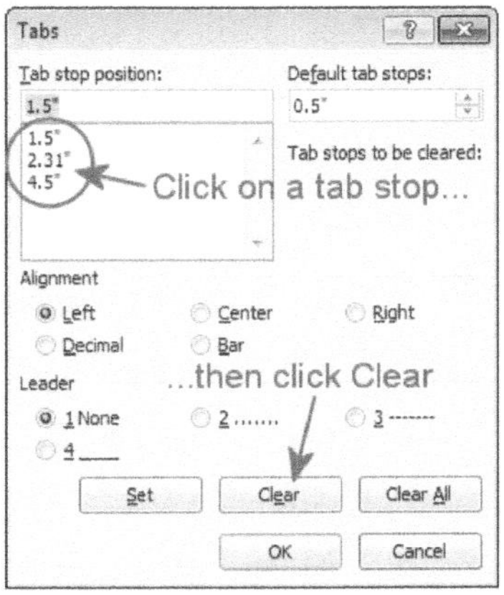

6. Click on Clear.
7. Repeat steps 5 and 6 for each tab stop you want to clear.
8. Click on OK.

If you have the ruler displayed on-screen (it appears just above your document window and just below the ribbon), you can also delete tabs by following these steps:

1. Make sure the insertion point is in the paragraph whose tabs you want to clear.
2. Use the horizontal scroll bar to scroll left or right until you can see the offending tab on the ruler.
3. On the ruler, click on the tab indicator and drag it off the ruler. When you release the mouse button, the tab should disappear from the ruler.
4. Repeat steps 2 and 3 for each tab stop you want to clear.

Adding a Tab at the Beginning of a Line

One annoying feature of Word is that the program assumes that if you press the **Tab** key at the beginning of a line, you want to indent the line. The program then proceeds to adjust the left margin by the distance of a normal tab stop. Although this setting can be helpful in some circumstances, I can think of many times when I really want the tab character inserted into the document instead of the margin information adjusted.

If you want to make sure that Word inserts a tab character, simply press **Ctrl+Tab**. This technique works at any time in Word, but it is most useful at the beginning of lines. As an alternative, you can turn off the indent feature by following these steps:

1. Display the File tab of the ribbon and then click Options. Word displays the Word Options dialog box.
2. At the left side of the dialog box click Proofing.
3. Click AutoCorrect Options button. Word displays the AutoCorrect dialog box.
4. Make sure the AutoFormat As You Type tab is displayed.

5. Make sure the Set Left- and First-indent with Tabs and Backspaces check box is not selected.
6. Click on OK.

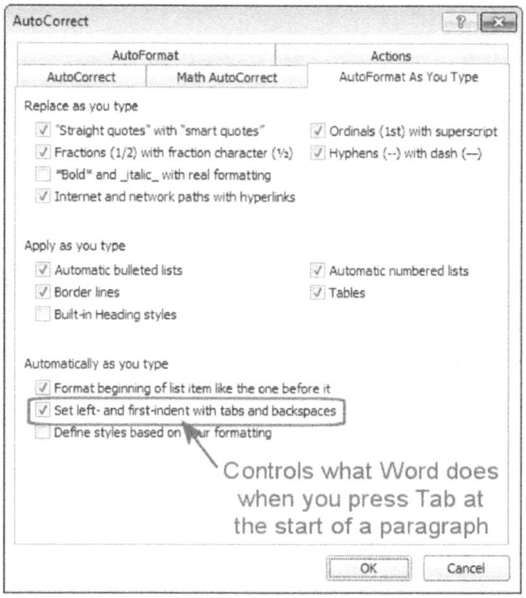

Understanding Default Tab Stops

Word has many different levels of formatting that you can do in a document. Some formatting settings apply to characters, others to paragraphs, still others to sections, and finally there are formatting settings that apply to the entire document.

We normally think of tab stops as being a "paragraph-level setting." However, there is one tab stop option that applies to the entire document, not just to a particular paragraph. The way in which the setting is presented in Word is a bit deceiving, however.

The setting in mind, of course, is for the default tab stop distance. There is a single default tab stop distance that you can set in Word, although you do it from the Tabs dialog box, which is definitely a paragraph-level formatting dialog box. To see what this means, position your insertion point anywhere within your document; it doesn't matter where. Display the Home tab of the ribbon and click the small icon at the lower-right of the Paragraph group. Word displays the Paragraph dialog box and you should click the Tabs button, at the lower-left of the dialog box. This displays the Tabs dialog box.

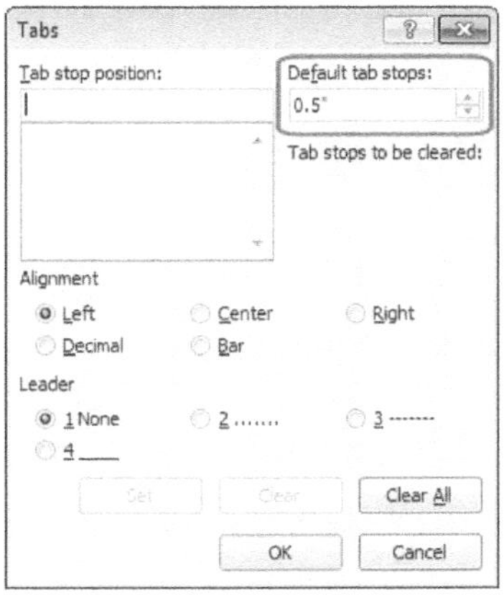

In the upper-right corner of the dialog box is the Default Tab Stops value. This value applies to the entire document, even though every other setting in the dialog box applies only to the current paragraph.

To test this out, change the Default Tab Stops value to something different from what it is, then click on OK. The dialog box closes, and the default is changed. Now, position the insertion point somewhere else in the document and again display the Tabs dialog box. The Default Tab Stops value should still be set to the new value you entered.

This means that to permanently change (so to speak) the default tab stop value, you need to do it within the templates you use to create new documents. For instance, you would follow these general steps to change the default tab stops in the Normal.dotx template:

1. Load the Normal.dotx template.
2. Display the Home tab of the ribbon.
3. Click the small icon at the lower-right of the Paragraph group. Word displays the Paragraph dialog box.
4. Click the Tabs button, at the lower-left of the dialog box. Word displays the Tabs dialog box.
5. Change the Default Tab Stops value on the dialog box.
6. Click on OK.
7. Save the template.

Once these steps are complete, then all new documents based on the Normal. dotx template will have the default tab stops set the way you want them. (Notice that this is for all new documents; existing documents will still have the old default tab setting. They have to be changed individually.)

Working with Columns

Planning on publishing a newspaper? Too bold, you say? How about publishing a newsletter, then? Either way, you need columns! Actually, columns aren't used only in laying out newspapers and newsletters. Word allows you to set up columns to your heart's content. In the following sections, you learn how to add columns, change the number of columns, put lines between columns, and change the width Word assigns to each column.

Adding Columns to Your Page Layout

As you format your documents in Word, you may find it better to lay out your text in columns. This is usually done if you are developing a newsletter or magazine layout. The number of columns you use is up to you—Word allows you to divide your page into as many as 100 columns, although this arrangement may look a bit strange. To add columns, follow these steps:

1. Position the insertion point at the beginning of the material you want to appear in columns.
2. Choose the Page Layout tab for the ribbon.
3. In the Page Setup group, click the Columns drop-down list.
4. Pick the number of columns you want to use for the selected text.

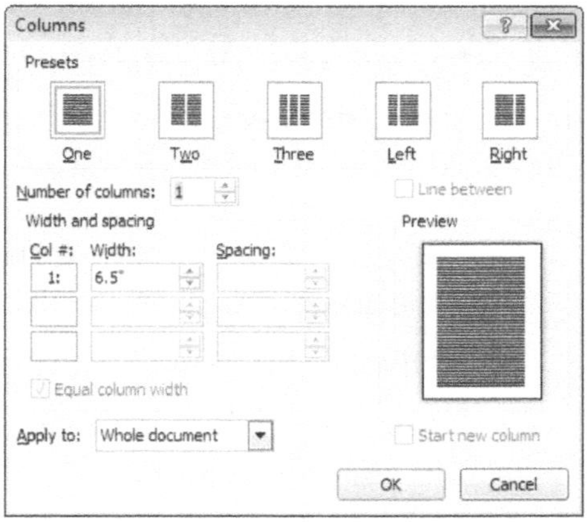

When you do step 3 you can select the most common numbers of columns (1, 2, or 3) or two different semi-fancy column layouts. If you want more columns then these, click the More Columns option in step 4 and you will see the Columns dialog box where you can set up to 45 columns.

There is an interesting side note here: If you use the up and down arrows in the Number of Columns field of the Columns dialog box to set columns, you can only choose a range between 1 and 13 columns when you are using portrait-oriented letter-size paper. If you change to landscape-oriented paper, then you can use the control to go between 1 and 18 columns. It appears that Word sets the upper limit of the control so that no individual column on the paper will be less than a half-inch wide. If you change the printable area of your page (adjust the paper size, orientation, and margins), then Word adjusts the upper limit of the Number of Columns control.

Don't let this fool you, however. You can still set up to 45 columns—sort of. You just have to type the number into the Number of Columns control. If you try to type a number that is larger than 45, Word generates an error message that says you can only set between 1 and 45 columns. So, you can try to type a larger number into the control, but when you do and then start working with a different control in the dialog box, Word adjusts the number of columns back down to the calculated maximum discussed previously. (Word just won't let you set a column width narrower than half an inch, and this limits the number of columns you can get on the page.)

What if you have a document and you want to format part of it in columns? As an example, let's assume you have a five-page document, and you want to format the center part of page 2 as three columns. You want rest of the document to remain a single column. To overcome this formatting challenge there are only two changes you need to make to the above steps. First, in step 1, you need to select the text that will appear in the columns. Second, you need to make sure you use the Apply To drop-down list to choose Selected Text.

Changing the Number of Columns

If you have already added columns to your document, Word lets you easily change the number of columns you are using. To do this, follow these steps:

1. If your document is formatted using more than one section, position the insertion point in the document section that you want to affect.
2. Display the Page Layout tab of the ribbon.
3. Click the Columns tab in the Page Setup group. Word displays a number of column configurations you can choose.
4. If you see the number of columns you want to use, click that option and you are done.
5. If you don't see the number of columns you want to use, click More Columns. Word displays the Columns dialog box.
6. Change the value in the Number of Columns field to the number of columns desired.
7. In the Apply To box, select Whole Document or This Section (depending on what you want your change to affect).
8. Click on OK.

Adding Vertical Lines Between Columns

As you are learning, Word allows you to add columns to your page layout. As part of your page layout, you may decide to add vertical lines between each column on your page. To do this, follow these steps:

1. Position the insertion point at the beginning of the material you want to appear in columns.
2. Display the Page Layout tab of the ribbon.
3. Click the Columns tool and then click More Columns. Word displays the Columns dialog box.
4. Specify the number of columns desired, as you normally would.
5. In the Apply To box, select This Point Forward. This causes Word to apply columns beginning where the insertion point is located.
6. Click on the Line Between check box.
7. Click on OK.

Changing the Column Width

After you add columns to your document, you may decide that you want to adjust the width of the columns. Word allows you to quickly adjust the width of individual columns in your document as well as the space between columns. You can change the width of columns in your document in the following manner:

1. Position the insertion point somewhere within the text that is already formatted for columns.
2. Display the Page Layout tab of the ribbon.
3. Click the Columns tool in the Page Setup group. Word displays a drop-down list of options.
4. Choose More Columns. Word displays the Columns dialog box.
5. Clear the Equal Column Width check box if you don't want all your columns to be an equal width. (The Equal Column Width check box is only accessible if you set the Number of Columns control to some value other than 1.)

6. In the Width and Spacing area of the dialog box, adjust the width and spacing (the space between columns) for each column to reflect your desired page appearance. As you adjust one number, other columns and widths may be adjusted automatically. All these numbers are interrelated.
7. Click on OK.

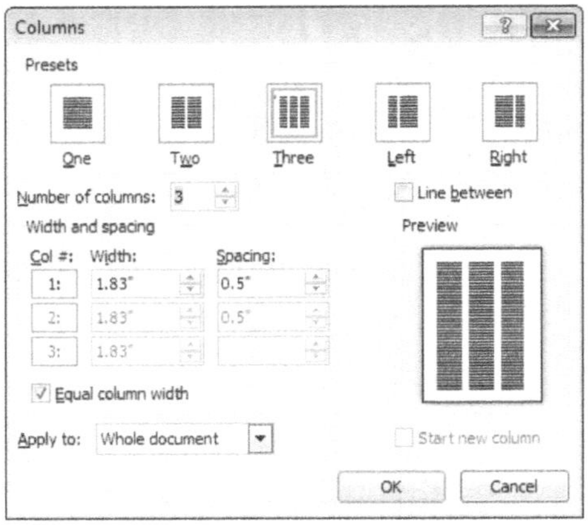

Working with Line Numbers

Many types of documents require the use of line numbers. Typically, they are used for legal documents. For instance, each line of a legal contract may need to be numbered. The following sections explain how to add, format, position, and turn off line numbers. You will even learn how to set a starting line number other than 1.

Adding Line Numbers

If you are creating a document that needs line numbers, Word allows you to easily add them. You do this in the following manner:

1. Position the insertion point anywhere in the section that you want to have line numbers. (Line numbering can be controlled on a section-by-section basis.)
2. Display the Page Layout tab of the ribbon.
3. Click anywhere on the Line Numbers tool (in the Page Setup group) to display a drop-down list of options.
4. Select Line Numbering Options. Word displays the Layout tab of the Page Setup dialog box.
5. Click on the Line Numbers button. Word displays the Line Numbers dialog box.
6. Make sure the Add Line Numbering check box is selected.

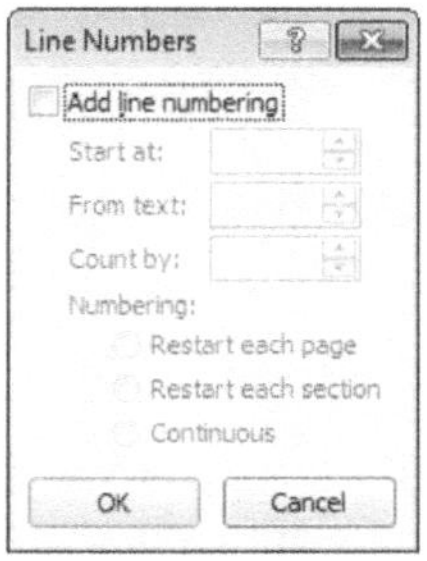

7. In the Line Numbers dialog box, specify the number at which you want to begin the numbering and the other characteristics you want used.
8. Click OK to close the Line Numbers dialog box.
9. Click OK to close the Page Setup dialog box.

You should note that you can see line numbers only when working in Print Layout view. If you are working in one of the other views, the line numbers are hidden. They will, however, appear on printouts, regardless of the view you are using.

Formatting Line Numbers

Word allows you to add line numbers to your documents, as you learned in the preceding section. You might be wondering how to change the formatting of those line numbers. Word makes this process quite simple by relying on styles for its line numbers. Simply follow these steps:

1. Display the Home tab of the ribbon.
2. Click the small icon at the bottom-right of the Styles group. Word displays the Styles pane at the right side of the desktop.

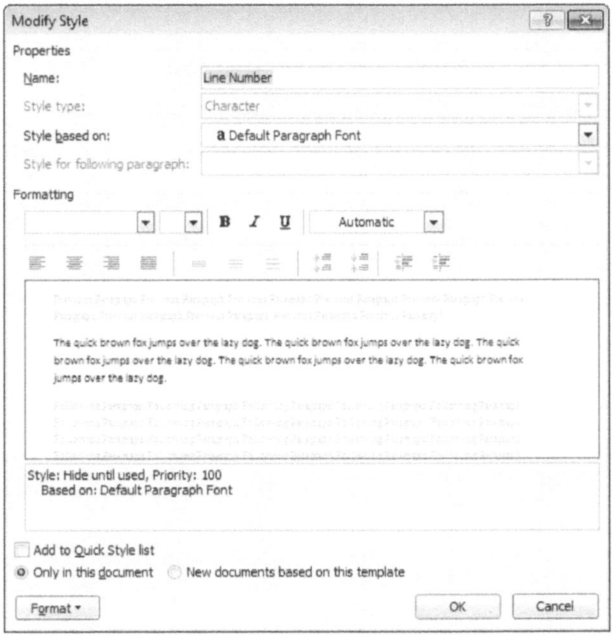

3. Hover the mouse pointer over the Line Number style. Notice that a downward-pointing arrow appears at the right side of the style name. (If the Line Number style is not visible, use the Options link at the bottom of the Styles pane to change which styles are displayed in the pane.)

4. Click on the downward-pointing arrow and choose the Modify option from the resulting options. Word displays the Modify Style dialog box.
5. Click on Format and choose the portion of the style you want to change. The options available through the Format button utilize dialog boxes with which you are already familiar.
6. Click on OK to save your modifications.
7. Close the Styles pane, if desired.

Setting the Starting Line Number

If you are working on a very long document and you have parts of the document stored in different files, you may not want the line numbers to start over at 1 for each file you print. To adjust the starting line number for a document, follow these steps:

1. Display the Page Layout tab of the ribbon.
2. Click the Line Numbers tool (in the Page Setup group) and then choose Line Numbering Options. Word displays the Layout tab of the Page Setup dialog box.

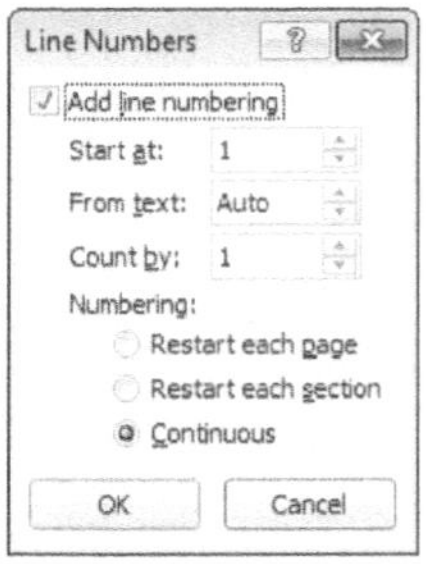

3. Click on the Line Numbers button. Word displays the Line Numbers dialog box.
4. Adjust the starting line number by changing the number in the Start At box.
5. Change the line number increment value (Count By) if desired.
6. Click on OK to close the Line Numbers dialog box.
7. Click on OK to close the Page Setup dialog box.

Positioning Line Numbers

In an earlier section, you learned how to add line numbers to your document. Word also allows you to modify how far line numbers are from the text in your document. To do this, follow these steps:

1. Display the Page Layout tab of the ribbon.
2. Click the Line Numbers tool (in the Page Setup group) and then choose Line Numbering Options. Word displays the Layout tab of the Page Setup dialog box.
3. Click on the Line Numbers button. Word displays the Line Numbers dialog box.
4. Adjust the measurement in the From Text box to indicate how far the line numbers appear from your text.
5. Click on OK to close the Line Numbers dialog box.
6. Click on OK to close the Page Setup dialog box.

Turning Off Line Numbering Temporarily

In the last few sections, you learned how to turn on and use the line-numbering feature. (It works great for legal or research-oriented documents.) At times, however, you may want to exclude a certain paragraph from line numbering. You do this in this manner:

1. Position the insertion point in the paragraph that you don't want to have line numbers.
2. Display the Home tab of the ribbon.
3. Click the small icon at the lower-right corner of the Paragraph group. Word displays the Paragraph dialog box.
4. Make sure the Line and Page Breaks tab is selected.
5. Select the Suppress Line Numbers check box.
6. Click on OK.

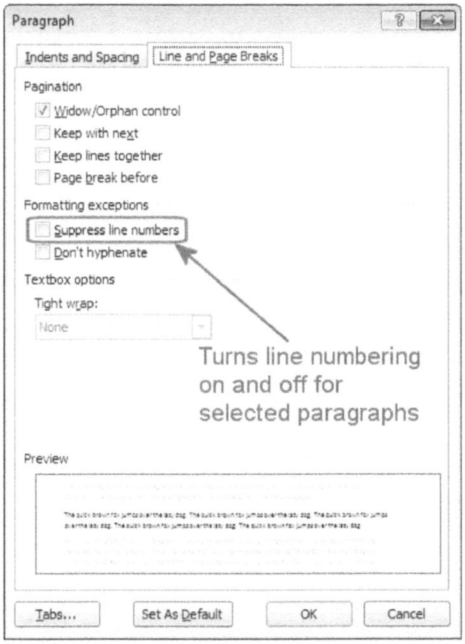

Introducing Lists

Word allows you to easily create both bulleted lists and numbered lists. In the following sections, you learn more about lists—how to create them, change them, and remove them.

You should realize that the topic of lists, in Word lingo, can get quite involved. For this reason, I could have spent quite a bit of time discussing lists in this book. I purposely haven't done that, however, because I want instead to provide only an introduction to the topic. If you need more information, make sure you consult a more advanced Word book.

Understanding and Creating Lists

Two types of lists are commonly used in printed material. The first is a bulleted list, and the second is a numbered list. A ***bulleted*** list is nothing but a list of related items with a symbol to the left side of the first line of each item in the list. For example, the following is a bulleted list:

- This is the first item in the list.
- This is the second item in the list. More than one line is in this item. Notice that the extra lines are aligned with the first line, not with the bullet or the text margin.
- This is the third item in the list.
- This is the fourth item in the list.

In the case of this bulleted list, the symbol used as the bullet is a small dot. A numbered list is a little different. It consists of a series of items, each with a sequential number in front of it. Both numbered lists and bulleted lists are used extensively throughout this book; you can find many examples of each.

This usage points out the primary way you can decide which type of list to use. If you have a sequence of steps that must be followed in order, you should use a numbered list. If you have a group of items to which you want special treatment given, but they don't represent a series that must be followed in sequence, you should use a bulleted list.

Word allows you to quickly and easily create lists. To do so, follow these steps:

1. Type your list and press **Enter** only at the end of each item in the list. If an item runs more than one line, do not press **Enter** at the end of each line.
2. Select all the items (paragraphs) in the list.
3. Display the Home tab of the ribbon.
4. Click on the Bullets tool to create the bulleted list or on the Numbering tool to create a numbered list. (Both tools are in the Paragraph group.)

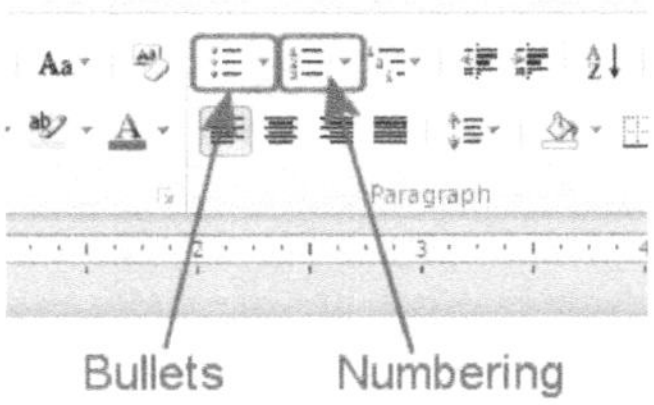

Changing the Types of Numbers in a Numbered List

When you are working with numbered lists, Word uses Arabic numerals, by default, to number the items. This number is followed by a period. You can change the numbering scheme Word uses to number your lists. Any of the following numbering types can be selected:

Type	Example
Arabic	1, 2, 3, 4
Uppercase roman	I, II, III, IV
Lowercase roman	i, ii, iii, iv
Uppercase alphabetic	A, B, C, D
Lowercase alphabetic	a, b, c, d
Ordinals	1st, 2nd, 3rd, 4th
Word numbers	One, Two, Three, Four
Word ordinals	First, Second, Third, Fourth
Leading-zero Arabic	01, 02, 03, 04

You can also change the type of character that follows the numbered list number. You can even define characters to appear in front of the number. To change the numbering type and the character following the number, follow these steps:

1. Select the paragraphs you want included in your numbered list.
2. Display the Home tab of the ribbon.
3. Click the down-arrow next to the Numbering tool in the Paragraph group. Word displays the Numbering Library.
4. Click Define New Number Format. Word displays the Define New Number Format dialog box.
5. In the Number Style drop-down list, select a numbering format.
6. In the Number Format box, add any other text you want besides the actual number.
7. Modify the position of the number and the text, if desired, by using the Alignment control.
8. Click on OK to close the Define New Number Format dialog box.

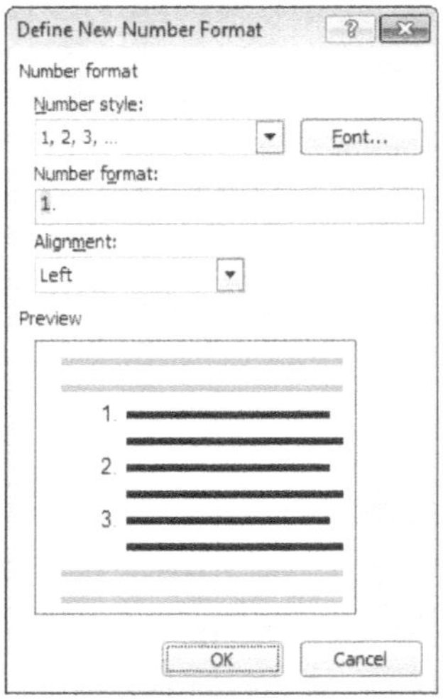

Removing a List

Sometimes, you may want to remove the list formatting you previously applied to some of your text. For instance, you may have a numbered list that you want to reformat as regular text. Word provides an easy way for you to remove list formatting:

1. Select the list items you want to no longer be a part of the list. It doesn't matter whether the list is a numbered or bulleted list.
2. Display the Home tab of the ribbon.
3. If the list is a numbered list, click on the Numbering tool in the Paragraph group.
4. If the list is a bulleted list, click on the Bullets tool in the Paragraph group.

If you remove the bullets or numbering from only a portion of the list, the remainder of the list retains the list formatting. If required, Word adjusts the numbering on any remaining part of a numbered list.

6

Introducing Styles and Templates

Unlike most other word processing programs, Word does an excellent job of separating content from appearance. This statement may sound strange, but the words you type are your content, and the way they look on-screen or on paper is their appearance. ***Styles*** are nothing more than a named definition of how text should appear. You can best understand this concept by comparing your text to water (this is your content). The appearance of the water depends on the attributes of the container in which it is placed. If you place it in a glass, it looks one way; if you place it in a pitcher, it looks a different way. The relationship between text and styles is no different: If you change the style that is applied to text, the appearance of the text automatically changes.

Although Word allows you to format your text explicitly by selecting it and then picking the attributes you want applied (as discussed in the preceding chapter), styles give you much more power. The reason is that you need to define the style only once and then you can apply it to text as you see fit. Plus, if you later change the style, all text formatted with that style is automatically updated to reflect the change. (You have changed the container, so the water's appearance changes.)

Templates, on the other hand, allow you to save your styles together with other things, such as macros. You can later apply all the elements of the template to a document just by attaching the template to the document.

In this chapter, you are introduced to the power and sophistication of both styles and templates. Here, you learn everything you need in order to take your formatting skills to the next powerful level.

Working with Styles

To many people, the concept of working with styles seems complex—something they want to avoid. Working with styles is not that difficult, however, after you understand the basics. In the following sections, you learn those basics: what types of styles you can create, how to define them, how to apply them, and how to change them. You will also learn how to manage styles by renaming and deleting them.

Understanding Styles

At the beginning of this chapter, you learned what styles are—they allow you to save a collection of formatting attributes under a name you define. You can later apply that style to text, and the text will then adopt the attributes defined in the style. Word 2010 allows you to define five types of styles:

- Character. Character styles are used to define how individual characters appear, including attributes such as font and font size and bold, italics, and superscript.
- Paragraph. Paragraph styles define not only how the characters in the paragraph appear, but also how the paragraph should appear in relation to the margins of your document, whether it should include bullets or numbering, how it should be treated by the spelling and grammar checkers, and how it should appear in relation to other paragraphs in the document.
- Linked. The linked style is essentially a combination of both character and paragraph styles. When a linked style is applied to a text selection, it behaves just like a character style. When a linked style is applied to a paragraph of text, it behaves just like a paragraph style. Most of the built-in styles in Word are defined as linked styles.
- Table. Table styles are used to specify how a particular table appears, including borders, spacing, and other table-specific formatting attributes.
- List. List styles are used to define how bulleted lists and numbered lists should appear.

Styles are saved with a document, or they can be stored in a document template. Word allows you to add, change, rename, and delete styles, as you will shortly learn.

Defining a Style

You can define styles in a number of ways, but the most common is to use the Styles task pane. To define a style, simply follow these steps:

1. Display the Home tab of the ribbon.
2. Click the small icon at the lower-right corner of the Styles group. Word displays the Styles task pane at the right side of the document.
3. Click on the New Style button in the lower-left corner of the Styles task pane. Word displays the Create New Style from Formatting dialog box. You use this dialog box to set the attributes of the style.

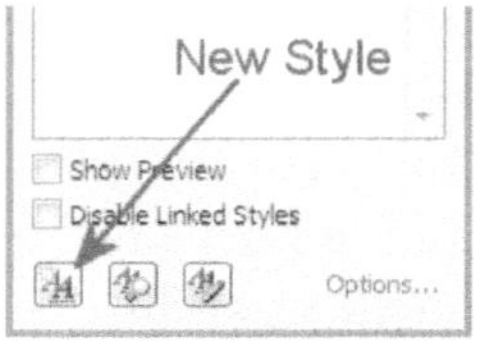

4. Make sure you specify the name and type of style you are creating. (There are five types of styles, as discussed in the previous section.)
5. Use the Style Based On drop-down list to indicate if this new style is based on (derived from) an existing style.
6. You can make changes to common attributes directly in the New Style dialog box. If you need to make more detailed attribute changes, click on the Format button. The types of formatting available through the Format button depend on the type of style you are defining.

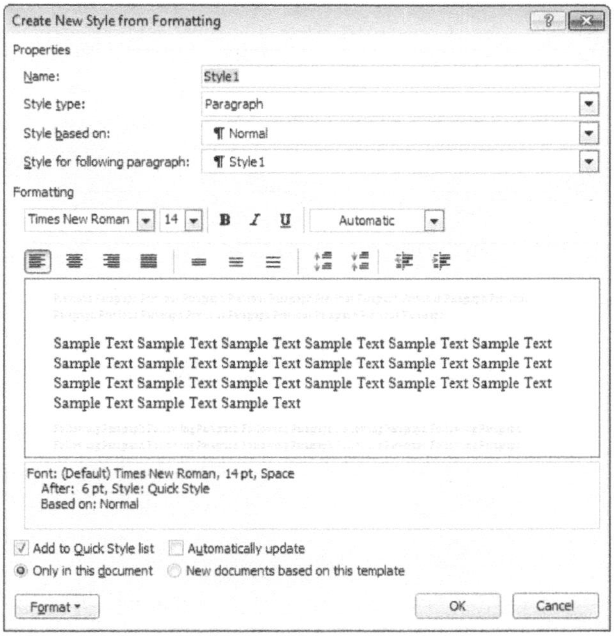

7. When you are done setting the formatting attributes, click OK to close the dialog box. Your newly defined style should appear in the formatting list in the Styles task pane.
8. Close the Styles task pane, if desired.

At this point, you can use your style anywhere you like within your document.

Intelligently Starting a New Paragraph

If you use styles in your documents, you can use a powerful Word feature. The Next Style feature, which is defined when you set up a style, indicates what Word should use as the style following this one. Assume that you have two styles defined: Figure and Caption. The Figure style is used for figures in your document and is always followed by the Caption style, which is used for the figure caption. You can set the Next Style feature for the Figure style so that when you press **Enter**, Word automatically makes the next paragraph a Caption style.

To use the Next Style feature, follow these steps:

1. Display the Home tab of the ribbon and then click the small icon at the bottom-right of the Styles group. Word displays the Styles task pane.
2. In the list of styles shown, hover the mouse pointer over the name of the style you want to change. A drop-down arrow should appear at the right side of the style name.
3. Click the drop-down arrow and choose Modify. Word displays the Modify Style dialog box.
4. In the Style For Following Paragraph drop-down list, select an existing style that you want to follow this style (when **Enter** is pressed). In the above example, this would be the Caption style.
5. Click on OK.
6. When you are done, close the Styles task pane if desired.

When modifying a style, you may want to select the radio button called New Documents Based On This Template. This saves the changes you are making in the template file for use in other documents.

Applying a Style

Styles do you no good if you cannot apply them to your text. You can apply styles in three ways.

The first way depends on your use of the ribbon. On the Home tab of the ribbon you see the Styles group. It contains a series of styles that are known as a ***Style Set.*** You can click any of these styles and that style is applied to whatever text is selected in your document. (You can also hover the mouse pointer over any of these styles and the text in the document momentarily changes to reflect the style over which you are hovering. This allows you to see what a style looks like before actually applying it by clicking.)

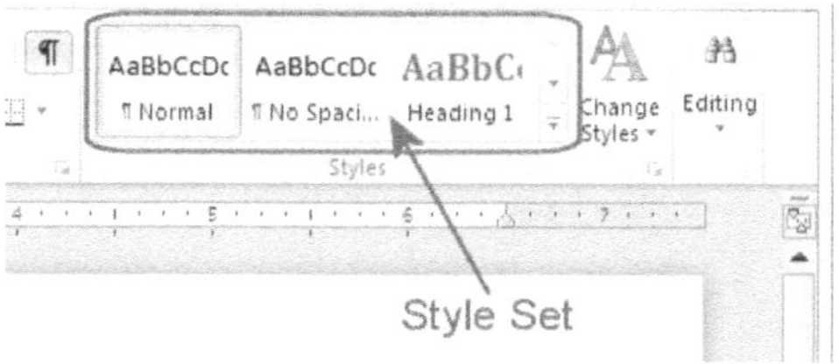

The second method of applying styles uses the Styles task pane. You can follow these steps to apply a style:

1. Select the portion of your document you want to format. For instance, position the insertion point in the paragraph to which you want to apply a paragraph style, or select the characters to which you want to apply a character style.
2. Click the small down-pointing arrow icon at the lower-right corner of the Styles group on the Home tab of the ribbon. Word displays the Styles task pane.
3. In the list of styles displayed, click the name of the style you want to apply.

The third method assumes that you have assigned your styles to specific shortcut keys. If you have, just select what you want to format and then use the shortcut key.

The three most common types of styles applied in a document are character, paragraph, and linked styles. Word also allows you to define table and list styles, but they are not used as often as character, paragraph, and linked styles. As their names imply, character styles define how individual characters should appear, while paragraph styles are more comprehensive and define how entire paragraphs should appear. Linked styles act as either character or paragraph styles, depending on what you apply them to.

Word includes quite a number of built-in styles you can use for your documents. Some of these are defined as character styles, some as paragraph styles, but the majority are linked styles. If you examine the Styles task pane, you can tell what type of styles each style is:

- Character styles have a lowercase 'a' at the right of the style name
- Paragraph styles have a backwards P (a pilcrow character) at the right of the style name
- Linked styles have both a lowercase 'a' and a pilcrow at the right of the style name

The general rules by which style application is governed are as follows:

1. If characters are selected and you apply a character style, then the attributes of that style are applied to the selected characters.
2. If characters are selected and you apply a paragraph style, then the attributes of that style are applied to the entire paragraph, not just the selected characters.
3. If characters are selected and you apply a linked style, only the character attributes of the selected style are applied. The original style used for the overall paragraph format remains unchanged.
4. If no characters are selected and you apply a character style, then there is no effect on the paragraph. If you immediately start typing, however, the character style determines the appearance of what you type.
5. If no characters are selected and you apply either a paragraph style or a linked style, then the style changes for the entire paragraph and it assumes the attributes of the style you selected. If there were any characters in the paragraph that had individual formatting applied (through styles or through manual formatting), then that formatting remains intact.
6. If an entire paragraph is selected (as characterized by including the paragraph mark in your selection) and you apply a character style,

then all the characters in the paragraph assume the attributes of the selected style.

7. If an entire paragraph is selected and you apply either a paragraph style or a linked style, then the formatting for the entire paragraph is changed. If there were any characters in the paragraph that had individual formatting applied, then that formatting remains intact.

Changing a Style

After you have defined a style in Word, you can change it at any time. To do this, follow these steps:

1. Display the Home tab of the ribbon and then click the small icon at the bottom-right of the Styles group. Word displays the Styles task pane.
2. From the list of styles displayed in the Styles task pane, locate the style you want to change and simply point to it. (Don't click on it.) Word displays a down-arrow at the right side of the style name.
3. Click on the down-arrow. Word displays a drop-down list of options.
4. Click on Modify. Word displays the Modify Style dialog box.
5. Change the attributes of the style, as desired. You may need to use the Format button to make your changes.
6. When completed, click on OK.
7. To change other styles, repeat steps 2 through 6.

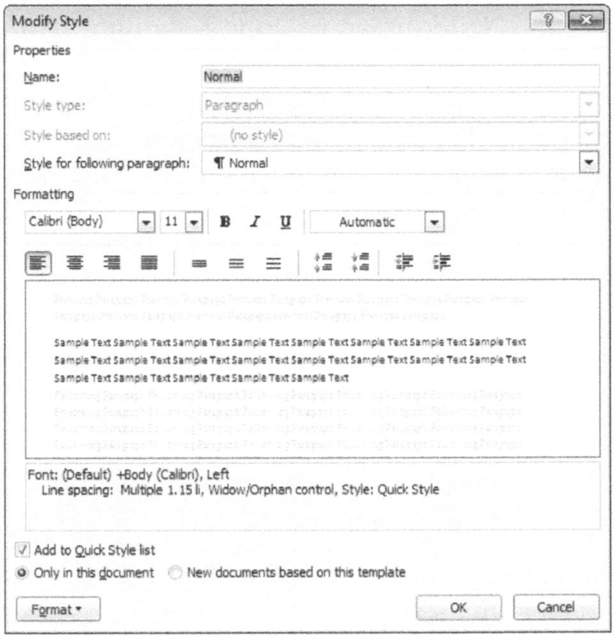

8. When you are done changing styles, close the Styles task pane.

Renaming a Style

Word allows you to rename styles you have defined. You cannot, however, rename any of the predefined styles supplied with Word. To rename a style, follow these steps:

1. Display the Home tab of the ribbon and then click the small icon at the bottom-right of the Styles group. Word displays the Styles task pane.
2. Scroll through the styles in the Styles task pane until you see the style you want to rename.
3. Hover your mouse pointer over the style name. You should see a drop-down arrow appear at the right side of the style name.
4. Click the drop-down arrow and choose Modify. Word displays the Modify Style dialog box.
5. In the Name box, enter a new name for the style.
6. Click on OK.
7. To rename other styles, repeat steps 2 through 6.
8. When you are done renaming styles, dismiss the Styles task pane.

You can rename any style you want. If you rename a style you created, the name is changed as you directed. If you change the name of one of Word's built-in styles, Word simply adds your new name as a suffix to the original name of the style.

Deleting a Style

As you work with styles in Word, you may eventually want to delete a style you previously defined. To do this, follow these steps:

1. Display the Home tab of the ribbon and then click the small icon at the bottom-right of the Styles group. Word displays the Styles task pane.
2. In the list of styles shown in the pane, hover the mouse pointer over the style you want to delete. A down-arrow appears at the right side of the style name.

3. Click on the down-arrow that appeared in step 2. Word presents you with a list of actions you can take.
4. Choose either the Revert To Base Style option or the Delete option from the list of actions. You will be asked to confirm you action; click on Yes.
5. To delete other styles, repeat steps 2 through 4.
6. When you are done, close the Styles task pane.

You cannot delete styles that are predefined in Word. If you select these styles, the Delete option is not available in step 3.

Note in step 4 that there are two possible options you can use to actually delete the style. If the style is based on an underlying style (see the section entitled "Defining a Style," paying particular attention to step 5) then you can only "revert" the style. This means that all the text that uses the style you are deleting is changed to the underlying style for the style you are deleting. If there is no underlying style, then you have the option to actually delete the style.

Adding an Individual Style to the Template

Templates are used in Word to define the default styles you want used with a document or family of documents. (Templates are defined in more detail later in this chapter.) Word allows you to update the template when you either add or change existing styles. To do this, follow these steps:

1. Make sure the Home tab of the ribbon is displayed.
2. Click the small arrow at the bottom-right corner of the Styles group. The Styles pane appears at the right of the program window.
3. In the Styles pane, scroll through the list of styles until you see the style you want to change.
4. Move the mouse pointer over the style name and then click on the down-arrow at the right side of the style name.
5. Click on Modify. The Modify Style dialog box appears.

6. Make sure the New Documents Based On This Template radio button is selected.
7. Make any changes to the style you desire.
8. Click on OK.
9. To add or update other styles in the template file, repeat steps 3 through 8.
10. When you have finished changing styles, close the Styles pane.

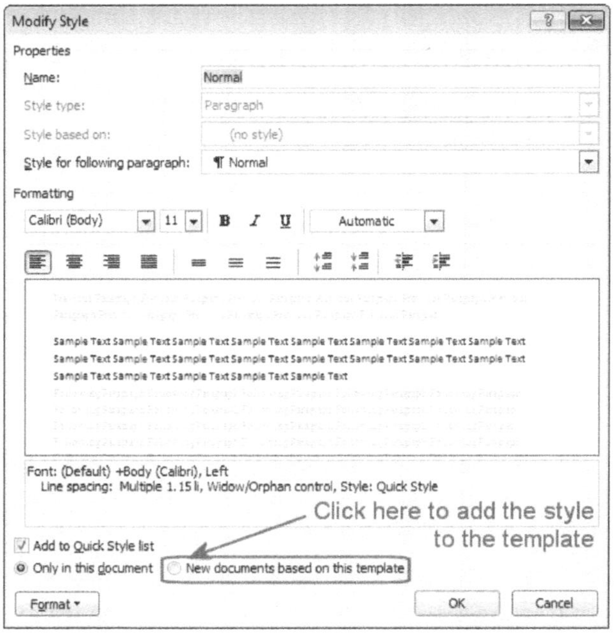

One Change Affects Everything

Perhaps you've had this happen to you: You are typing along and you decide you need to make a formatting change in your document. In this particular case, you want to format a particular paragraph as bold face. You select the paragraph, click on the Bold tool on the Home tab of the ribbon, and all the paragraphs in your document change to bold—not just the one you selected.

Or, you decide you want to add a bullet to the start of one paragraph. You select the paragraph, click the Bullets tool on the Home tab of the ribbon, and bullets appear in front of all the paragraphs in the document—not just the one you selected. What is going on, and how can this be corrected?

When this type of change occurs, you can press **Ctrl+Z** right away (the Undo shortcut) and Word reverses its action. Continually pressing **Ctrl+Z** is a pain, however. It is much better to understand the cause of the problem, then you can make the changes you need.

Every paragraph in your document is based on an underlying style that defines how that paragraph appears on the page. With this tidbit firmly understood, the next time you get a "global change" when you only wanted to affect a single paragraph, follow these four steps to greater enlightenment:

1. Press **Ctrl+Z** to undo the global change.
2. With the insertion point still in the paragraph you formatted (or the paragraph completely selected), display the Home tab of the ribbon and click the small arrow icon at the bottom right of the Styles group. Word displays the Styles task pane.
3. Hover the mouse pointer over the selected style and click the down-arrow at its right side.
4. Click on Modify. Word displays the Modify Style dialog box.

Take a good look at the dialog box. At the bottom is a check box labeled Automatically Update. If this check box is selected, it means that whenever you make changes to the paragraph, in your document, those changes are automatically made to the style that is assigned to the paragraph. Once the change is made, then every other paragraph in your document that is formatted using the same style is automatically changed.

In most cases, Word uses the Normal style as the default style for paragraphs in a document. If the Normal style has the Automatically Update check box selected, it is guaranteed that you will experience the problems described at the beginning of this tip.

To solve this problem, follow these steps:

1. Display the Modify Style dialog box, as described in the previous four steps.
2. Make sure the Automatically Update check box is cleared.
3. Select the radio button called New Documents Based On This Template.
4. Click on OK to close the Modify Style dialog box.
5. Close the Styles task pane.

Printing a Style Sheet

If you use styles within your documents, Word allows you to print a style sheet that you can use as a reference when applying and planning styles for your document. To print a style sheet, follow these steps:

1. Press **CTRL+P**. Word displays the File tab of the ribbon with the printing options visible.
2. Click the first button under the Settings label.
3. Click the Styles option.
4. Click Print.

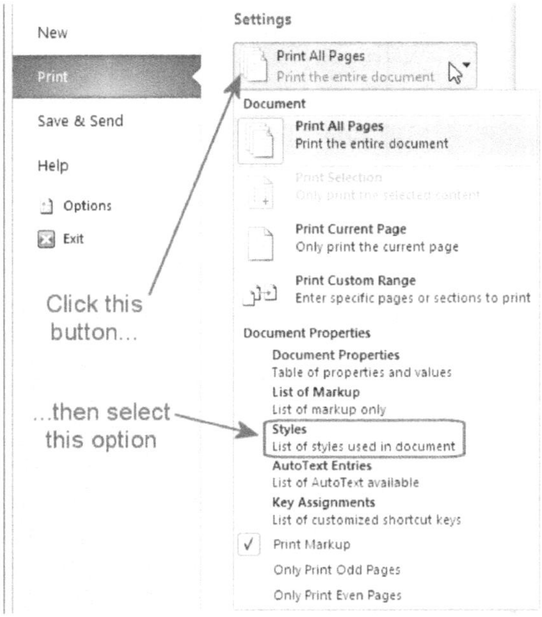

Using the Style Area

Word can show you the style names for all paragraphs in your document. This capability is useful if you are editing a document and need to see whether the appropriate styles have been applied. To reveal styles on the left side of the screen, follow these steps:

1. Display the File tab of the ribbon and then click Options. Word displays the Word Options dialog box.
2. At the left side of the dialog box click Advanced.
3. Scroll through the options until you see the Display section.
4. Set the Style Area Pane

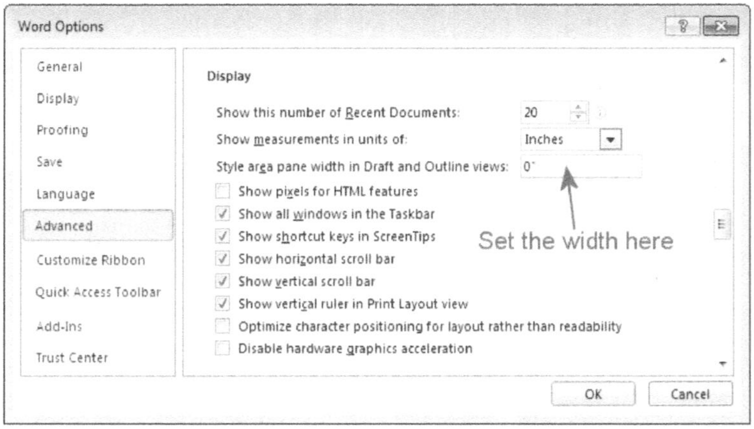

Width in Draft and Outline Views field to a value other than 0. (A good starting point is .5, which represents a half-inch.)

5. Click on OK to close the dialog box.

Now switch to Draft view. At the left side of your document you should now see a style area. Notice that the style assigned to each paragraph is now visible within the style area. If you want to change the width of the style area at this point, you can do so by using the mouse. Simply click on the vertical line that sets off the style area and drag it left or right.

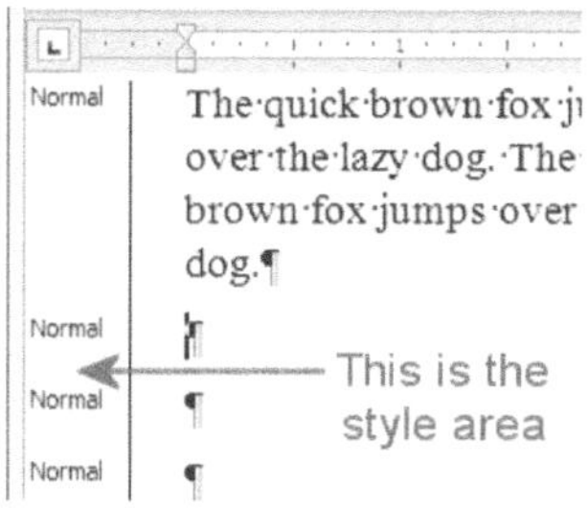

You should note that the style area is only visible if you are using Draft or Outline views. If you switch to any other view, the style area disappears.

Avoid Using the Normal Style

If you use styles extensively in your Word documents, you should consider not using the Normal style. This pre-defined style is used as the default for paragraphs. If you create different styles and apply them to all the paragraphs in your document, then you can later search for any paragraphs defined with the Normal style. These paragraphs have not been formatted properly (since you haven't applied any of your defined styles to them) and deserve your attention. When no more Normal paragraphs are in your document, you know that your formatting job is complete.

Use Linked Styles with Care

Earlier in this chapter you learned that the three most commonly used styles are character styles, paragraph styles, and linked styles. Linked styles allow you to define a single style that can be applied to either characters or paragraphs, depending on what you have selected when you apply the style.

Basically, if you have characters selected when you apply a linked style, then the style acts like a character style. If you have nothing selected or you have entire paragraphs selected, then it acts like a paragraph style.

This may sound very powerful, and it really is. For this reason you should use linked styles with care. For most purposes I suggest not using linked styles. When you create your own styles, make sure that they are either character or paragraph styles—stay away from defining linked styles. Why? Because linked styles act like chameleons, and they can unnecessarily complicate your formatting tasks. When you apply character or paragraph formatting, there is no question what you are applying, and your formatting is cleaner in the long run.

If you decide that you don't want to use linked styles, you can essentially turn them off in Word. You do this by displaying the Styles task pane (display the Home tab of the ribbon and click the small icon at the bottom-right of the Styles group) and selecting the Disable Linked Styles check box at the bottom of the pane. This doesn't hide any linked styles, but it causes linked styles to always act like paragraph styles when applying formatting.

Developing a Style Family

When creating formatting styles in Word, you can base one style on another style. This concept has a great application: You can develop style families. To implement this feature, you should develop a few base styles. For instance, you may have a base style for regular text, one for numbered lists, and one for headings. (In fact, you can use Word's predefined styles for these base styles.) After the base styles are defined, you can base other styles on the base styles. That way, if you need to make a change that affects an entire style family, you can change the base style, and all the styles based on it will change.

When you name your styles (both the base styles and all the other styles based on them), you may want to name the styles so that the families are grouped together. One way of doing this is to start each style belonging to the same family with the same character or two. For example, each style in your regular text family could start with the characters ***RT,*** each of the styles in your headings family could start with ***HD,*** and so on. In this way, all the related styles will be displayed together in any style list.

Working with Templates

After you have the concept of styles tucked firmly under your belt, you are ready to start working with templates. Effectively used, templates stop you from reinventing the wheel at the start of each new document.

In the following sections, you learn what templates are, where they are stored, how to edit them, and how to apply them to a document.

Understanding Templates

A *template* is nothing more than a pattern used by Word to describe how your document looks and how you interact with the document. Templates contain formatting information, a style sheet, macros, and boilerplate text.

Since a template serves as a pattern for a document, it can be very powerful. For instance, you might define a template for each type of document that you work with. If you do a weekly departmental report, you might define a template that serves as a pattern for the report. If you write letters, you could define two different templates—one for business letters and another for personal letters. The uses to which templates can be put are literally endless.

Word allows you to create as many template files as you desire. The names of template files end with a filename extension that designates the type of template: either DOTX (a template that does not contain macros) or DOTM (a template that contains macros). If you want to create a template, follow these steps:

1. Start a new document.
2. Make your desired changes to the formatting and macros.
3. Press **F12** to display the Save As dialog box.
4. Using the Save As Type list, choose Word Template or Word Macro-Enabled Template, depending on whether your template contains macros or not.

That's all there is to it. Generally, selecting a type of template in the Save As Type list will move you to your templates directory. However, you should make sure that you save your new template file in the same directory with all your other template files. That way the file will be available for future use.

Finding Where Templates Are Stored

It is unfortunate that Microsoft does not make it easier to modify templates. The first step in modifying templates is to load one, which means that you need to know where they are stored. Regrettably, the average user hasn't a clue where they are stored. If you want to know where your templates are stored, follow these steps:

1. Display the File tab of the ribbon and then click Options. Word displays the Word Options dialog box.
2. At the left side of the dialog box click Advanced.
3. Scroll through the list of options until you see the General section.
4. Click the File Locations button. Word displays the File Locations dialog box.

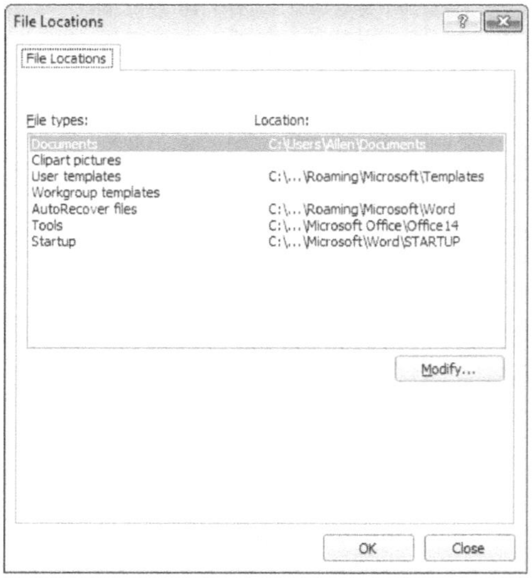

5. In the File Types list, choose User Templates. If the path for the templates is short enough, you may be able to see it in the dialog box right now. If so, you can skip steps 6 and 7.
6. Click the Modify button (even though you won't be modifying anything.). Word displays the Modify Location dialog box.
7. The Look In drop-down list, at the top of the dialog box, contains the current path name used for templates.

Template Folder Location

If you are a long-time Word user but are new to Word 2010, it is easy to become a bit confused—it seems that everything has changed from the version you are familiar with. For instance, if you create a template in Word, you may be curious as to where Word stores it. You may notice that when you create the template and store it on disk, it may not appear in the New Document templates.

The location for templates in Word 2010 is the same as it has been in other versions of Word. (To see what Word considers the location, follow the steps in the preceding section.) There is a difference in Word 2010 in how this location is handled when compared to some earlier versions of Word, however. The difference is that Word 2010 doesn't default to that folder when you are saving a file as a template. The reason for this change is security; the templates folder is trusted, meaning that any macros stored there are trusted by Word. To avoid the chance of placing an unwanted macro into a trusted location, Word 2010 doesn't store templates, by default, in the templates folder used by the program.

Fortunately, the template folder is now a default "place" in the hierarchy shown at the left side of the Save As dialog box. Click Templates in the hierarchy and the file will be saved in the proper templates location (the same one you saw in the File Locations dialog box). When you later want to create a new document based on the template, in the New Document dialog box you can select My Templates at the left side of the dialog box and see the templates that you've saved.

Editing a Template

Editing a template is much like editing a regular document; the only difference is that the file is saved with a different filename extension than regular documents. To load an existing template so that you can edit it, do the following:

1. Display the File tab of the ribbon and then click Open. Word displays the Open dialog box.
2. At the bottom of the dialog box, just above the Open button, is a drop-down list where you can specify the type of files you want Word to list. Use the drop-down list to select either Word Templates or Word Macro-Enabled Templates, depending on which type you want to open.
3. Using the controls in the dialog box, browse through directories and disks drives as desired until the desired template files are listed. (If

desired, you can browse to the folder containing the majority of your templates—the one you learned about in the previous two sections.)

4. Select the desired document template.
5. Click the Open button.

You can now make any changes desired, and then save the template again. The changes affect any future documents you base on the template.

Saving a Template with a Preview

When you click on the File tab and click New, Word displays the New Document dialog box. Click My Templates at the left side of the dialog box and Word shows you the templates available for creating your new document. On the left side of the dialog box are the templates, and on the right side is a preview of what that template looks like.

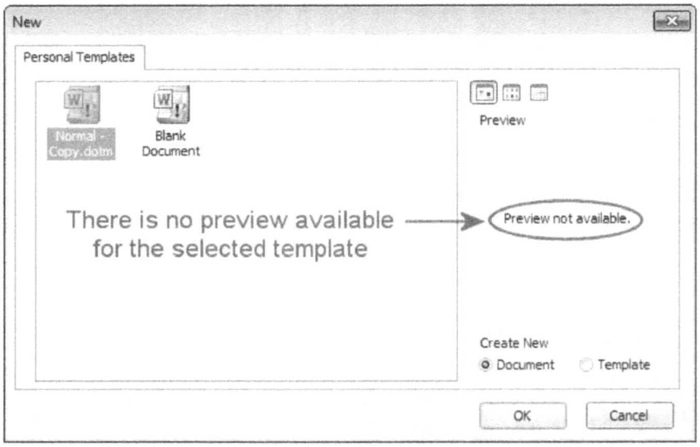

You may have already noticed that no preview is available in the New dialog box for some of your templates. If this is the case with your template and you want to have a preview available, you can follow these steps:

1. Display the File tab of the ribbon and then click Open. Word displays the Open dialog box.
2. At the bottom of the dialog box, just above the Open button, is a drop-down list where you can specify the type of files you want Word to list. Use the drop-down list to select either Word Templates or Word Macro-Enabled Templates, depending on which type you want to open.
3. Use the controls in the dialog box to locate and open the template file for which you want the preview saved.

4. Select the desired document template.
5. Click the Open button. Word opens the template.
6. With the template loaded, display the File tab of the ribbon.
7. Make sure the Info option is selected at the left side of the screen.
8. Click the Properties option at the right side of the screen. Word displays a few options you can choose.
9. Choose Advanced Properties. Word displays the Properties dialog box for your template file.
10. Make sure the Summary tab is selected.
11. Select the Save Thumbnails for All Word Documents check box at the bottom of the dialog box.
12. Click on OK to close the Properties dialog box.
13. Click the Save tool on the Quick Access Toolbar to save the template.

At this point, you can close the template file.

Now the preview should appear in the dialog box from which you select templates.

Changing the Attached Template

When you create a blank document, Word automatically attaches the Normal.dotx template to it, although you can specify a different template. After the file has been created, you can change the attached template by following these steps:

1. Display the Developer tab of the ribbon and then click on the Document Template tool. Word displays the Templates and Add-Ins dialog box.
2. Click on the Attach button. Word displays the Attach Template dialog box, which looks very much like a standard Open dialog box.
3. Use the controls in the dialog box to locate and select the template you want attached to your document.

4. Click on Open. The Attach Template dialog box disappears and the name of the template you selected appears in the Document Template box.
5. Select the Automatically Update Document Styles check box to make sure that the styles in the template are applied to your document.
6. Click on OK.

Note that step 1 requires you to display the Developer tab of the ribbon. If you can't find the Developer tab on your system, it is because you haven't instructed Word to display it. To uncover the Developer tab, follow these steps:

1. Display the File tab of the ribbon and then click Options. Word displays the Word Options dialog box.
2. At the left side of the dialog box click Customize Ribbon.

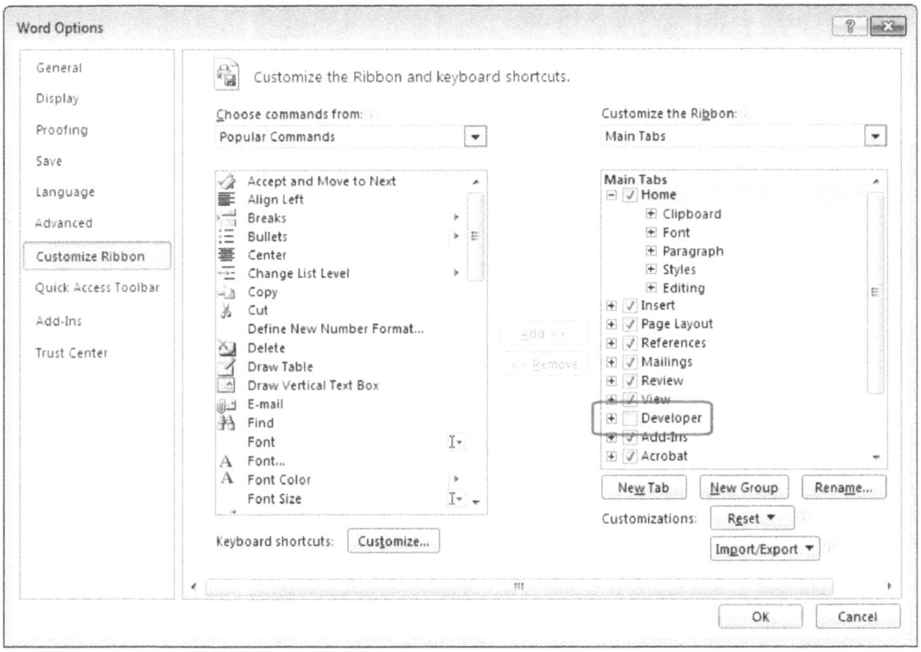

3. At the right side of the dialog box make sure that the Developer tab is selected in the list of tabs.
4. Click OK.

7

Using Tables

In the days of typewriters (and in some basic word processing programs), tables were created by using tabs to line up information. This method works just fine if your tables are not terribly complex or if you don't have multiple lines of text in each table cell. (A table ***cell*** is defined by the intersection of a row and a column.) In Word, you can use tables to format all your tabular data.

The table editor used by Word is an integral part of the program itself. It allows you to quickly and easily manipulate tables and the information contained in them. This chapter introduces you to the power of tables. Here, you learn how to create and modify tables, format tables, and edit information contained in tables.

Creating and Modifying Tables

The first step in using tables is to create your own. Word makes this process extremely easy, using a variety of different methods. In the following sections, you discover that you can insert, draw, or convert text into a table. You also learn how to merge, split, insert, and delete cells. You can even add rows and columns as desired. You even learn how to move entire rows of information around in your table.

Inserting a Table

There are several ways you can create a table in your document. Perhaps the easiest way to create a table is simply to insert the table. Follow these steps:

1. Place the insertion point at the location in your document where you want to insert the table.
2. Make sure the Insert tab of the ribbon is displayed.
3. Within the Tables group click the Table tool. Word displays a drop-down list of choices.
4. In the grid of blocks at the top of the drop-down list, click on the cell that illustrates the number of rows and columns you want in your table.

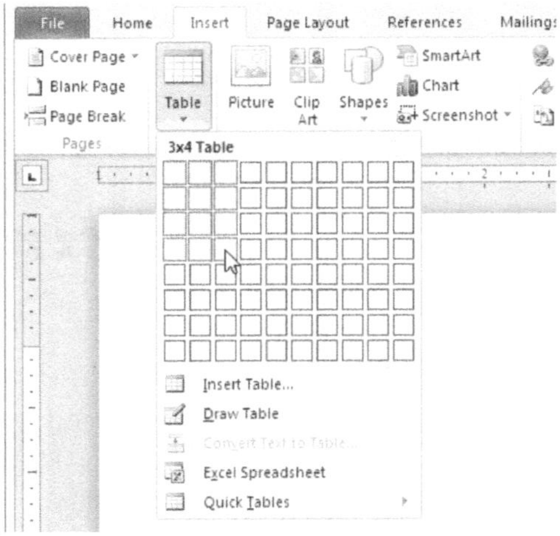

As you are inserting a table it actually appears in your document as you move your mouse over the cells in the drop-down list. Technically, the table is only actually inserted when you click a cell (step 4). If you don't click or if you abort the insertion by pressing **Esc**, then the table is never inserted and there is no change to your document.

Another way to insert a table involves using a dialog box. Choose the Insert tab of the ribbon, click the Table tool, and then choose Insert Table from the resulting options. Word displays the Insert Table dialog box, in which you can specify the number of columns and rows you want your table to have. Use the controls in the dialog box to set up your table, and then click on OK. Word inserts the table in your document.

Drawing a Table

As an alternative to using the methods described in the previous section is to use Word's draw-a-table feature. Follow these steps:

1. Display the Insert tab of the ribbon.
2. Click the Table tool on the ribbon. Word displays a drop-down list.

3. Select Draw Table from the list of options. Word changes to Print Layout view (if you are not already in that view) and changes the mouse pointer so it looks like a pencil.
4. Use the mouse pointer to define the outside borders of your table, much as you would draw in a drawing program.
5. Use the pencil to draw the columns and rows into the table.
6. Press **Esc** when you are done. Word changes the mouse pointer back to normal.

Converting Text into a Table

Besides creating a blank table (as you learned about in the preceding two sections), you can also convert existing text into a table. This is handy if you receive information that someone has formatted a table using tabs. To convert text into a table, follow these steps:

1. Make sure the text that is to be converted into the table contains only a single tab character between each column.
2. Select the text you want converted into a table.
3. Display the Insert tab of the ribbon.
4. Click the Table tool and then click Convert Text to Table. Word displays the Convert Text to Table dialog box.
5. Make sure all the table settings are correct and then click on OK.

The selected text is immediately converted into a table. In some instances, the width of columns in the resulting table may need to be adjusted. You will know if this is the case because the table will look very strange. Sometimes, you can get files from programs, such as spreadsheets, that use commas to separate columns. Word also converts this comma-separated text to a table.

Merging Table Cells

Word contains a powerful table editor that allows you to create complex tables. One feature of the table editor is that you can merge adjacent cells together. ***Merging*** cells simply means that the adjacent cells are thereafter treated as a single cell, even though they are not really a single cell. If the cells are on the same row, you can merge them together by following these steps:

1. Select two or more adjacent cells, on the same row or same column, that you want to merge.
2. In the Table Tools Layout tab of the ribbon, in the Merge group, find and click on the Merge Cells tool.

The gridline between the two cells is eliminated, and the two cells are joined. You can also easily perform cell merging by using the Eraser tool:

1. In the Design tab of the ribbon (visible when the insertion point is in a table) look in the Draw Borders group and click on the Eraser tool. Word changes the mouse pointer into an eraser shape.
2. Clicking the mouse pointer on a line between cells will delete the line and merge the cells. Clicking and dragging the mouse pointer across lines will also delete them.
3. Click on the Eraser tool again (on the ribbon) or press the **Esc** key. This turns off the Eraser tool.

Splitting Table Cells

In the preceding section you learned how to merge cells within a table. After cells have been merged, you can later split them apart using the following steps:

1. Right-click the previously merged cell.
2. Choose Split Cells from the resulting Context menu. Word displays the Split Cells dialog box.

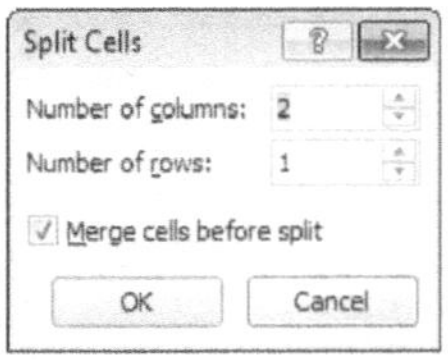

3. Using the controls in the dialog box, specify the number of columns and rows into which the merged cells should be split.
4. Click OK.

After you split the cells, the cell widths may be a bit different from the other cells in the table, and you may need to readjust them.

You can also split previously merged cells by using the tools on the ribbon in this manner:

1. Display the Design tab of the ribbon. (This tab is only visible if the insertion point is somewhere within a table.)
2. Click the Draw Table tool in the Draw Borders group. This is the one that looks like a pencil. The mouse cursor now looks like a pencil.
3. Use the mouse cursor to draw cell lines in your table. Simply click and drag to draw each new cell line. When you release the mouse button, the cells appear as you have drawn them.
4. When you are finished drawing, click on the Draw Table tool again or press the **Esc** key. This turns off the drawing mode.

Inserting a Cell

Sometimes you need to increase the size of your table. Word allows you to insert cells in your table by following these steps:

1. Select the cell before which you want a cell inserted.
2. On the Layout tab of the ribbon, click the small down-pointing arrow at the bottom-right of the Rows & Columns group. Word displays the Insert Cells dialog box.

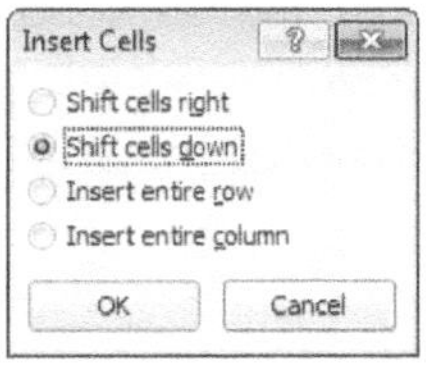

3. Select which way you want the cells to be adjusted.
4. Click on OK.

Deleting a Cell

Word contains a powerful table editor that allows you to create complex tables. You can delete individual cells in your table, even though such an action may make the table asymmetrical. To delete a cell, follow these steps:

1. Position the insertion point in the cell you want to delete.
2. Choose the Layout tab of the ribbon. (This tab is only available when you are working in a table.)
3. Click Delete from the Rows & Columns area of the ribbon, then choose Delete Cells from the resulting drop-down list of options. Word displays the Delete Cells dialog box.
4. Select which way you want the cells to be adjusted.
5. Click on OK.

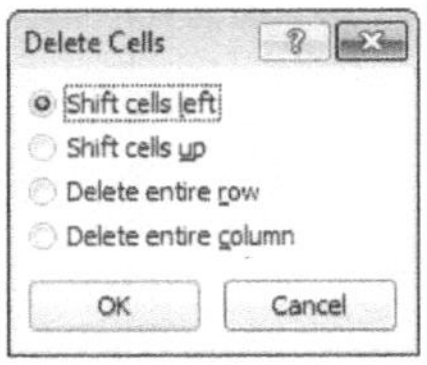

Splitting a Table

Word provides a powerful table editor that allows you to create and modify data in tabular format. If you are working with large tables, there may be times when you need to split a table in half. To do this, follow these steps:

1. Position the insertion point anywhere in the row you want as the first row in the table that is split-off the larger table.
2. Display the Layout tab of the ribbon. (This tab is only visible if your insertion point is within a table.)
3. Click the Split Table tool, within the Merge group.

The result is that Word inserts a Normal-formatted paragraph prior to the table row in which your insertion point was placed. The two resulting tables can now be treated independently.

Deleting a Table

There are several different ways you can delete a table. Here is one method:

1. Select the entire table you want to delete.
2. Display the Layout tab of the ribbon.
3. Click Delete (in the Rows & Columns group) and then click Delete Rows.

You can also use the Cut option from the Home tab of the ribbon to remove a table. Pressing **Del** will have no effect, however, other than to clear the contents of the cells in the table. (You should note that pressing **Del** will actually delete a table if you select at least one paragraph in addition to the table.)

You can also delete a table by following these steps:

1. Place the insertion point anywhere within your table.
2. Display the Layout tab of the ribbon.
3. Click Delete (in the Rows & Columns group) and then click Delete Table.

Quickly Inserting Table Rows

When you need to insert rows into a table, the easiest way is to follow these simple steps:

1. In the existing table, select a number of rows equal to the number you want to insert.
2. Display the Layout tab of the ribbon.
3. Click the Insert Above tool, in the Rows & Columns group. The rows are inserted just before the selected rows.

You can also, if preferred, click the Insert Below tool in step 2. This results in the new rows being added below whatever rows you selected in step 1.

Moving a Row or Column

Moving a row or column in a table is almost as easy as moving regular text. You do it by using the Clipboard. To move a row, follow these steps:

1. Select the row you want to move.
2. Press **Ctrl+X** or click on the Cut tool on the Home tab of the ribbon. This removes the row from the table and copies it to the Clipboard.
3. Move the insertion point to the beginning of the row before which you want to place the row you just cut.
4. Press **Ctrl+V** or click on the Paste tool on the Home tab of the ribbon.

Moving a column is a similar task and is accomplished by following these steps:

1. Select the column you want to move.
2. Press **Ctrl+X** or click on the Cut tool on the Home tab of the ribbon. This removes the column from the table and copies it to the Clipboard.
3. Move the insertion point to the beginning of the top cell of the column before which you want to place the column you just cut.
4. Press **Ctrl+V** or click on the Paste tool on the Home tab of the ribbon.

Moving a Table Row Quickly

In the preceding section, you learned how you can move table rows by using the Clipboard. You can also move rows by following these steps:

1. Position the insertion point in the row you want to move, or select the rows you want to move.
2. While holding down the **Shift** and **Alt** keys, press the **Up Arrow** or **Down Arrow** to move the row.

Note that when you are using this method, you can even move a row (or rows, because you can select more than one row in step 1) right out of the table. In other words, the moving doesn't stop when you reach the top or bottom of the table. As you continue pressing the **Up Arrow** or **Down Arrow** keys, the rows continue to move away from the table in the direction you choose.

This technique does not work to move columns, however. You should note that when you hold down the **Shift** and **Alt** keys and press the **Left Arrow** or **Right Arrow**, the heading level of the entire row is changed.

Moving a Row or Column with the Mouse

As long as you have Drag-and-Drop Text Editing turned on (in the advanced options of the Word Options dialog box), Word allows you to use the mouse for many routine editing tasks. One such task involves moving rows or columns within a table.

To move a row or column using the mouse, follow these steps:

1. Select the entire row or column that you want to move.
2. Click on the highlighted row or column, and hold down the mouse button. Shortly the pointer should change to a "ghost" insertion point with a small box next to the pointer arrow.

3. Drag the row or column to the place where you want it to be. You should position the insertion point in the row or column before which your moved row or column will appear.
4. Release the mouse button. The row or column is moved.

Copying a Row or Column with the Mouse

In the preceding section, you learned that you can move rows and columns by using the mouse. You can also use a similar technique to quickly copy rows and columns. Simply follow these steps:

1. Select the entire row or column you want to move.
2. Hold down the **Ctrl** key as you click on the highlighted row or column. Make sure you hold down the mouse button as well. Shortly, the pointer should change to a "ghost" insertion point with a small box next to the pointer arrow.
3. While still holding down the **Ctrl** key, drag the row or column to the place where you want it to be. You should position the insertion point in the row or column before which your copied row or column will appear.
4. Release the mouse button. The row or column is copied to that location.

You may notice something interesting about copying rows and columns in this manner: You can also "extend" the boundaries of your table. Granted, adding new rows or columns by copying them can, on its own, be viewed as extending the boundaries. However, let's say you have five rows in your table and you are copying one column using the mouse. If you release the column (step 4) in any place other than in the first row, the number of rows in your table is increased and the column is pasted in an "offset" manner. For instance, in your five-row table, if you release the column in row 3, the copied column is pasted beginning at row 3 and extending downward five rows, which means that you effectively add two new rows to your table, in addition to the expected new (copied) column.

You might think that the same "extension" would happen if you were moving rows. In a sense, it does, but with one important difference: If you release the row in some place other than in the first column, the new (copied) row—and only that row—extends past the right edge of the existing table. With columns, all the columns are extended into the new rows. With rows, only a single row is extended.

Converting a Table into Text

You may sometimes want to convert an existing table into text. To do this, follow these steps:

1. Select the entire table you want to convert to text.
2. Make sure the Layout tab of the ribbon is displayed.
3. In the Data group, click the Convert to Text tool. Word displays the Convert Table to Text dialog box.
4. Select the appropriate character that Word should use to separate the columns of text.
5. Click on OK.

Formatting Tables

When working with tables, you are not limited to a stodgy, spreadsheet-like appearance for your data. Instead, Word allows you to change and format virtually every aspect of how a table appears. In the following sections you learn how to resize columns, adjust row height, and change how information is aligned in cells. You also learn how to indent tables and resize the entire table. You even learn how to force Word to keep the contents of a row from splitting across a page break.

Resizing a Table Column with the Mouse

Word provides a variety of ways you can change the width of table columns using the mouse. The exact effect of each method depends on the version of Word you are using.

The simplest resizing you can do is to simply point to a column border with the mouse pointer and drag the border to a new location. This adjusts the widths of the columns separated by the border, but the overall width of the table remains the same.

If you hold down the **Shift** key as you drag a column border, the width of the column to the left of the border is changed, and the overall table width is adjusted accordingly. No other column widths are affected.

If you hold down the **Ctrl** key as you drag a column border, width of the column to the left of the column border is adjusted, and the overall table width remains the same. All columns to the right of the table border are adjusted so they are wider or narrower in a proportional value.

If you hold down both **Shift+Ctrl** as you drag a column border, the width of the column to the left is changed, the overall table width remains the same, and all column widths to the right are adjusted so they share equally in the remaining table width.

Holding down the **Alt** key as you drag a border adjusts columns the same as if you simply clicked and dragged a border. The only difference is that Word adds precise width notations in the Ruler area for each column. If you hold down **Alt+Shift**, you cannot drag any of the column borders. If you hold down **Alt+Ctrl** or **Alt+Shift+Ctrl**, the effect is the same as just holding down **Ctrl** (or **Shift+Ctrl**), with the addition of the precise measurements.

Adjusting the Column Width Using Ribbon Tools

In the preceding section, you learned how to adjust the width of table columns by using the mouse. If you want to adjust the size of table columns so that they are precise, you can't beat the tools on the ribbon. You can adjust the column width by following these steps:

1. Select the column whose width you want to change.
2. Display Layout tab of the ribbon. (This tab is visible only if you perform step 1.)
3. Click the Properties option in the Table group. Word displays the Table Properties dialog box.

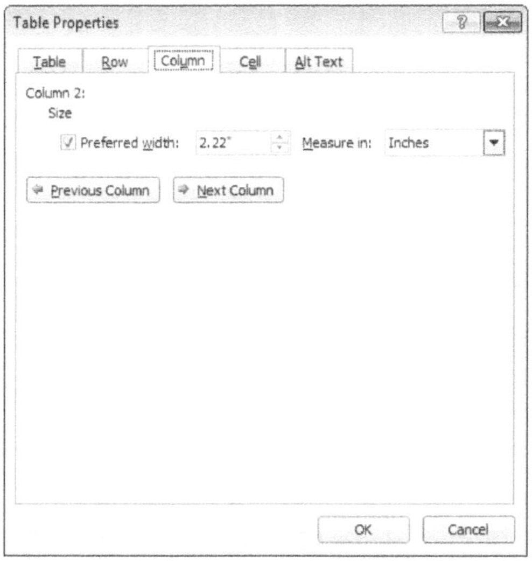

4. Make sure the Column tab is displayed.
5. Using the Preferred Width control, set the column width as desired.
6. Use the Previous Column or Next Column buttons, as desired, to select other columns.
7. Repeat steps 5 and 6 until you have set all the column widths desired.
8. Click on OK to close the Table Properties dialog box.

Resizing the Row Height with the Mouse

If you need to adjust the height of rows in a table, you can do so quickly and easily with a mouse. All you need to do is point to a row border with the mouse pointer and drag the border to a new location. Word does not affect the height of any row except the one just above the border you are dragging.

If you hold down the **ALT** key as you are adjusting row height, Word displays precise measurements in the vertical ruler area at the left side of the document. This technique is a great way to see the exact height of each row as you are making adjustments.

Adjusting the Table Row Height

Word allows you to independently adjust the height of every row in a table. For many table uses, you can rely on Word to select an automatic row height.

At other times, you may want to take charge and specify your own row height. You can do so by following these steps:

1. Select the table row whose height you want to specify.
2. Display the Layout tab of the ribbon.
3. In the Cell Size group, adjust the Height setting, as desired.
4. For more height options, click the Table Properties icon at the bottom-right of the Cell Size group. Word displays the Table Properties dialog box.
5. On the Row tab, specify a height and how you want that measurement interpreted (At Least or Exactly).
6. Click on OK.

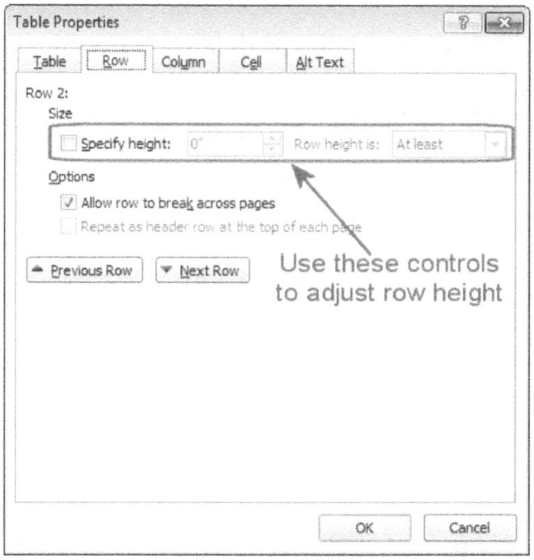

Changing the Cell Alignment

If you desire, you can independently specify the alignment of every cell in a Word table. Originally, the cell alignment matches the alignment of the paragraphs from which the table was created. You can override this, however, by using the paragraph alignment buttons in the Paragraph group of the Home tab of the ribbon. Simply position the insertion point in the cell whose alignment you want to adjust, and then select one of the buttons.

Centering Information in Table Cells

Tables are a powerful formatting tool used in many Word documents. Often, you must center information within the cells of a table. You can center information in two ways: horizontally and vertically.

Horizontal centering is quite simple, as you learned in the preceding section. Centering text vertically, however, is handled differently. All you need to do is follow these steps:

1. Right-click on the cell containing the information you want to vertically center. This displays a Context menu for the cell.
2. Choose the Cell Alignment option from the Context menu. This displays a submenu.
3. Choose one of the options that center the content vertically (there are three to choose from).

You should note that vertical centering will not give the desired results if you have the paragraph formatting for the text set to anything except single spacing, with no space before or after the paragraph. This extra spacing before or after affects the centering in the same way that paragraph indents affect horizontal spacing.

Hiding Gridlines

You may not realize it when you first insert a table, but Word includes some nonprinting gridlines around the table. You can't normally see the gridlines because they are obscured by the border that Word uses, by default, to format tables. The purpose of gridlines is to mark the boundaries of the table and each cell in the table when you have the table formatted for no border.

Table gridlines are turned off by default, but you have complete control over whether gridlines are displayed or not. You can turn gridlines on by placing the insertion point within a table and then displaying the Layout tab of the ribbon. In the Table group click the View Gridlines tool. When you later want to turn gridlines off, again follow the same steps.

The gridlines setting is not set on a per-table basis. In other words, if you turn the gridlines off anywhere in a document (in or out of a table), it is turned off throughout the entire document.

Designing Standard Tables

When you insert a table in Word, it adds a set border of one-half point around each cell in the table. If your "standard" table is different from what Word thinks it should be, however, you may be up a creek, since Word doesn't allow you to define what a standard table should look like.

If you have a "standard" table that you use over and over again in your document, there is a way around this problem. Remember, however, that a "standard" table (by this definition) will always have the same number of columns and rows and be exactly the same as every other standard table. (Of course, you can modify your standard table once it is inserted in your document.)

I've found that the easiest way to create your standard table is through the use of the building blocks that Word allows you to create. (I know that you can create table styles, but I find them less than easy at times.) All you need to do is follow these general steps:

1. Create your standard table, as desired. Make sure it is formatted as desired, and that it includes any standard text.
2. Select the entire table.
3. Press **Alt+F3**. The Create New Building Block dialog box appears.
4. Provide a name for your standard table in the Name field.
5. Using the Gallery drop-down list, choose Tables.
6. Click on OK.

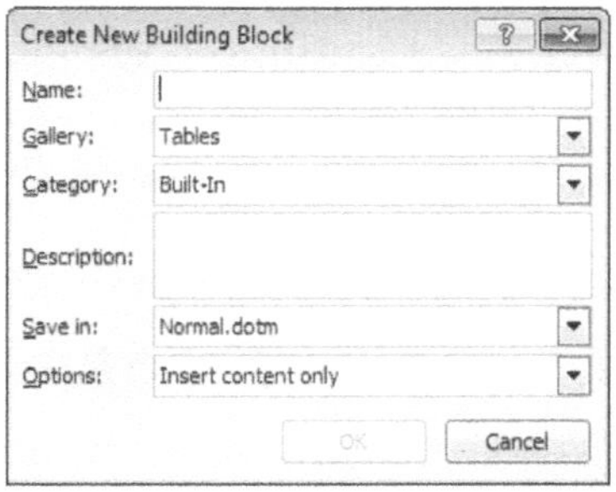

Your standard table is now created. To use the table, simply type the name you entered in step 4, and then press **F3**. The table appears in your document.

Changing the Text Direction in a Cell

Word allows you to quickly change the direction in which the text in a cell faces. For instance, you may want the text to be vertical instead of horizontal. You can change text direction by following these steps:

1. Right-click your mouse on the cell whose text direction you want to change. A Context menu appears.
2. Choose Text Direction from the Context menu. The Text Direction dialog box appears.

3. Select the Orientation you want used for your text.
4. When satisfied, click on OK.

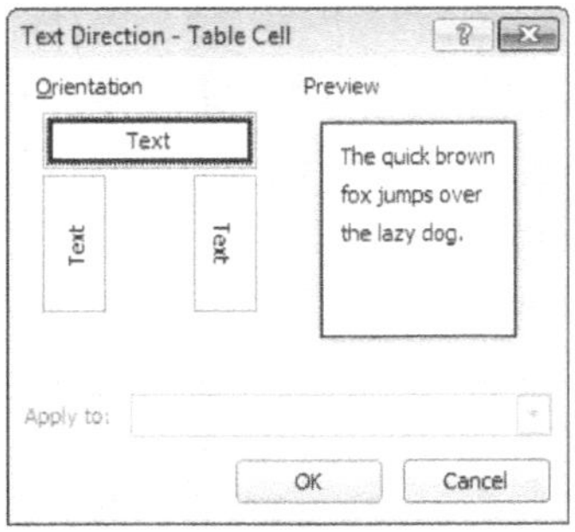

You can also do the same task using the Layout tab of the ribbon. Follow these steps:

1. Position the insertion point in the cell whose orientation you want to change.
2. Make sure that the Layout tab of the ribbon is displayed.
3. Click the Text Direction option in the Alignment group on the ribbon. Word changes the orientation of the text in the cell.
4. Continue clicking on the Text Direction option until the orientation is exactly as you want it.

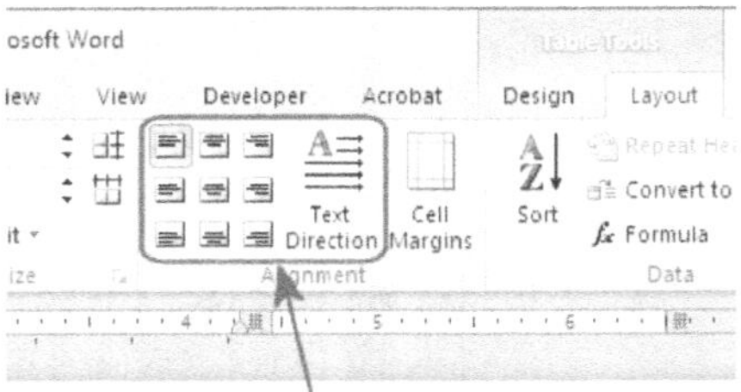

Text direction tools

Changing the Spacing Between Table Cells

Word contains a powerful table editor, which allows you to create complex tables. Normally, Word does not add any additional space between the cells in a table. You can manually adjust the amount of space, however. Follow these steps:

You can also

1. Make sure the insertion point is somewhere within your table.
2. Display the Layout tab of the ribbon. (This tab is only visible if you complete step 1.)
3. Click the Cell Margins tool in the Alignment group. Word displays the Table Options dialog box.
4. Make sure the Allow Spacing Between Cells check box is selected.

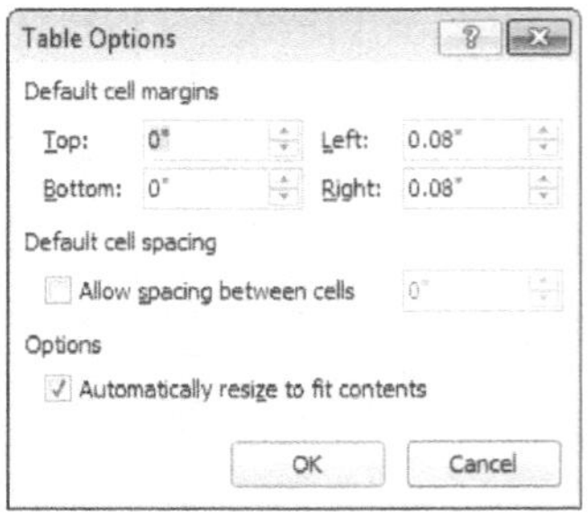

5. In the control to the right of the check box, specify how much space to leave between cells.
6. Click on OK.

Keeping Table Rows Together

You may believe that you can keep tables rows together if you select the text in a row and then choose Keep Lines Together from the Line and Page Breaks tab of the Paragraph dialog box. This is a common trick that you can read about in lots of places. What you may not know is that choosing this option may not always give the desired result.

The reason is that Word apparently ignores this setting within tables. (Astounding, isn't it?) Instead, Word pays attention to a setting in the Tables dialog box. If you look at the Row tab of the Table Properties dialog box, you see a check box entitled Allow Row to Break Across Pages. This is the only option that controls whether a page can break in the middle of a row. If the option is not set (the check box is clear), then the row won't break, regardless of the Keep Lines Together setting. Conversely, if the option is set, then the row can break, even if that means splitting up the paragraph text within the row.

Upon reflection, you might think there is method to this madness. After all, the setting in the Paragraph dialog box should only affect paragraphs, and the setting in the Tables dialog box should affect tables. Under this logic, however, you would expect that if a table cell contains a long paragraph (10 or 12 lines), and the page break is going to occur in the middle of the paragraph, that the Keep Lines Together setting would still keep the single paragraph together. Unfortunately, it doesn't. Word still pays attention to only the Allow Row to Break Across Pages setting.

Fitting Your Text in a Cell

Word has an odd little feature that allows you to force text to fit within a single line of a cell. The program does this by decreasing the apparent size of the text so that everything fits. Word does this (according to the Word on-line help system) without changing the actual font size of the text. It makes little real-world difference, however, because the size of your text is changed on both the screen and on any printouts.

It might be more accurate to say that although the height of the text that is used is not changed, the width is. You can take advantage of this Word table feature by following these steps:

1. Select the cell or cells that you want to format.
2. Make sure the Layout tab of the ribbon is displayed.
3. Click the small icon at the bottom-right corner of the Cell Size group. Excel displays the Table Properties dialog box.
4. Make sure the Cell tab selected.
5. Click on the Options button. Word displays the Cell Options dialog box.
6. Make sure the Fit Text check box is selected.
7. Click on OK to close the Cell Options dialog box.
8. Click on OK to close the Table Properties dialog box.

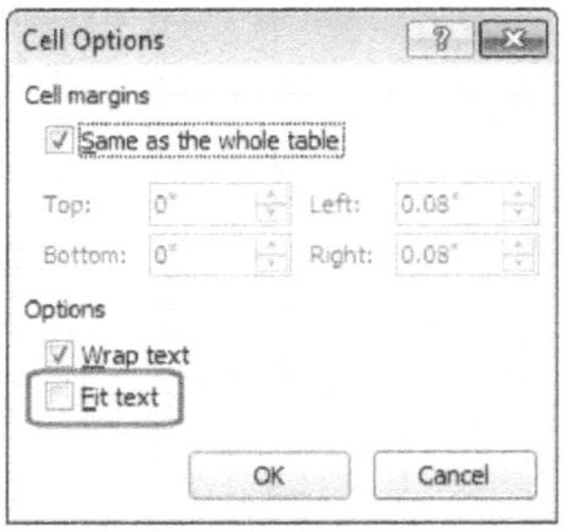

You should note that if a cell does not have enough text to fit all the way across the cell, Word forces the text to fit. Word does this by increasing the spaces between words and characters to fill the cell; it does not, however, widen the actual characters that make up the cell contents. Word apparently only makes characters narrower, not wider.

Indenting a Table

After you have inserted a table in your document, you cannot use the regular formatting tools to indent the table. This is because in most instances, tables do not behave the same way as paragraphs.

One way you can indent a table is to simply drag it to an indented position in your document. All you need to do is move the positioning marker for the table. This is the marker right above the leftmost edge of your table. When you move this marker, Word moves the entire table in the direction you move the marker.

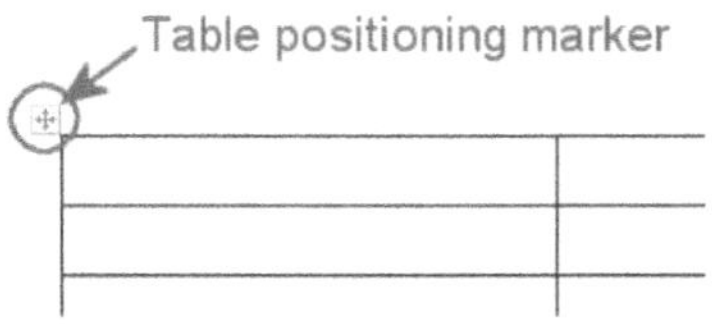

Another, more precise method of indenting your table is accomplished by following these steps:

1. Position the insertion pointer somewhere within the table you want to indent.
2. Make sure the Layout tab of the ribbon is displayed. (This tab is available only if the insertion point is inside a table, as you should have done in step 1.)
3. Click the Properties tool within the Table group. Word displays the Table Properties dialog box.
4. Make sure the Table tab is selected.
5. Using the Indent From Left control, specify how much of an indent to use for the table.
6. Click on OK.

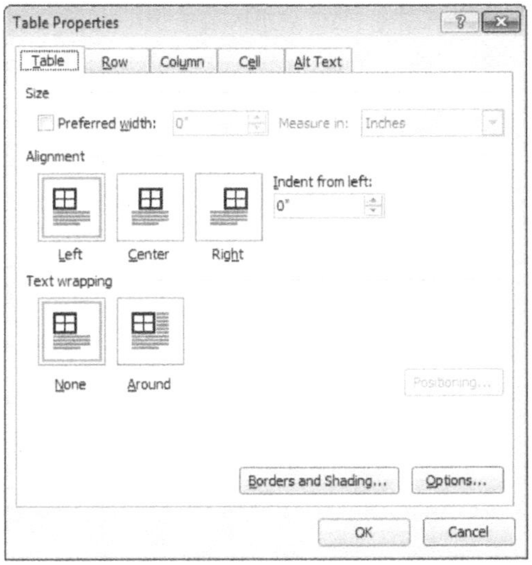

Aligning Decimal Numbers in a Table

When working in tables, you have no documented way of aligning decimal numbers. You can, if desired, add a decimal tab stop in a cell and then press

CTRL+TAB to align the number to the tab stop, but Word has an even easier way. All you need to do is follow these steps:

1. Select the cell or cells whose contents you want aligned.
2. Click repeatedly on the tab stop marker at the left edge of the ruler, stopping when you see the symbol for a decimal tab.

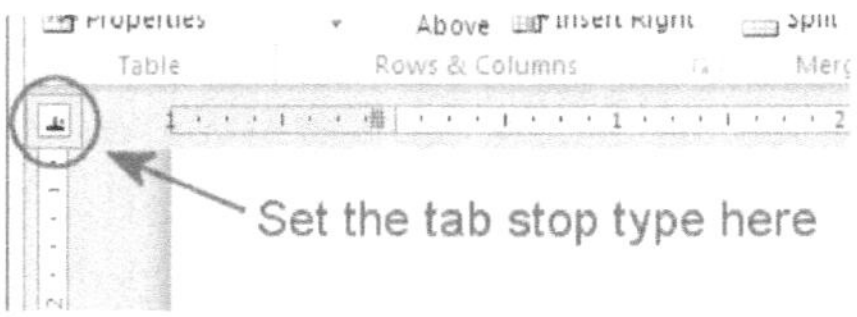

3. Click on the ruler above the selected cells, at the location where you want the numbers aligned.

Presto! The numbers get aligned as per the tab stop. You should note, however, that this is the only tab option (decimal) that works in this manner.

Spacing Table Rows Vertically

When you are working with tables, you may wonder how you can easily add space before a table row. This issue can be important when you have the table formatted with borders and the text in the table is too close to the border at the top of the row. Word makes it easy to add space after the information in a row, simply by adjusting the height of the row itself so that it is greater than the space occupied by the information in the row. But what about before a row?

An easy way to solve this problem is with paragraph formatting. If you have a single paragraph in each cell of a row, simply select the row. You can then follow these steps:

1. Display the Page Layout tab of the ribbon.
2. Adjust the value in the Spacing Before control (in the Paragraph group) so it represents the amount of space you want between the text and the row border. (A good setting is either 3 or 6 pt.)

You can also use this method if you have multiple paragraphs in each cell. The difference, however, is that you should select the top paragraph in each cell and format just it rather than format the paragraphs in the row as a whole.

Resizing a Table

One nifty shortcut in Word is the ability to resize your table using the mouse. You do this in much the same way as you resize graphics. All you need to do is position the mouse pointer over the table. Word displays a small, hollow square at the lower-right corner of the table. Click on this square, and drag it to represent the size you want your table to be. When you release the mouse button, the table is resized.

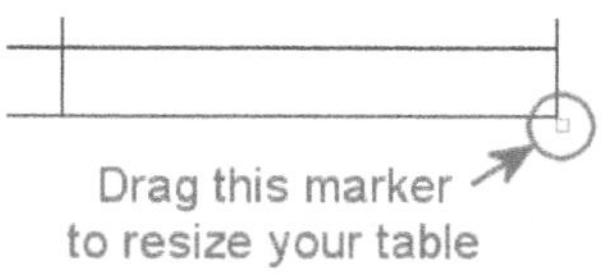

You should note that this technique does not change the number of columns or rows in your table. Instead, it simply adjusts the height of each row and width of each column so that the overall table size matches what you indicated you wanted.

Editing the Information in Tables

You already know how to edit the information in your documents. Editing regular text is a bit different from editing the information in a table, however. Sometimes, the ways in which you edit are similar, but there are enough differences to warrant the inclusion of the following sections.

In the following sections, you learn how to select tables, rows, columns, and cells. You also learn how to clear table contents and work with table headings.

Selecting a Table

There are some editing tasks that require you to select an entire table in Word. For instance, you may want to change the font used in the table, or you may want to adjust the borders around all the table cells. You probably already know that you can position the insertion point within the table and then choose Select Table from the Layout tab of the ribbon. This is obviously a two-step process, however, since it requires the positioning of the insertion point within the table before the table can be selected.

Another, perhaps easier way to select the table is to hover the mouse pointer over the table. You should notice a small square appear at the upper-left corner of the table. This square has a four-headed arrow within it. If you click on the square a single time, Word selects the entire table.

Selecting a Column or Row in a Table

You can select a column within a table in two ways:

- Position the insertion point in any cell within the column, display the Layout tab of the ribbon, click the Select tool, and then choose Select Column.
- Move the mouse cursor just above the column you want to select. It will turn into a downward pointing arrow. Click on the left mouse button.

The two ways to select a row within a table are similar:

- Position the insertion point in any cell within the row, display the Layout tab of the ribbon, click the Select tool, and then choose Select Row.
- Move the mouse cursor just to the left of the row you want to select, and then click on the left mouse button.

Selecting Individual Cells in a Table

Sometimes, you want to apply formatting to individual cells in a table, or you may need to perform some other action that requires selecting cells first. You can select a cell in a couple of ways. If you are using the mouse, you can select a cell by moving the mouse pointer just to the left of the cell itself. The mouse pointer should point up and to the right. When you click on the mouse button, the entire cell should be selected.

Another way to select cells is to simply select the end-of-cell marker at the end of any text in the cell. This process is especially easy if you are using the keyboard. All you need to do is make sure the insertion point is at the end of your text (in the cell) and then hold down the **SHIFT** key as you press the **RIGHT ARROW** key. The result is that the entire cell is selected. If you cannot see the end-of-cell markers, you can view them by instructing Word to display nonprinting characters.

Picking a Contiguous Range of Table Cells

Many times, when you are working with tables in Word, you need to select a range of cells. This task is easy to do using the keyboard, the mouse, or a combination of both.

The first step in selecting a range of cells is to place the insertion point in the first cell in the range. With this part done, you are ready to proceed:

- Using the keyboard, you can hold down the **SHIFT** key as you use the navigation keys (up, down, left, and right arrows) to extend the range.
- Using the mouse, you can click on the first cell and hold down the mouse button as you drag the mouse to extend the range.
- Using the keyboard and mouse, you can hold down the Shift key as you click on the cell that defines the opposite corner of the range you want selected.

Clearing the Contents of a Table

If you spend lots of time working on a table and getting it just right, you may want to copy the table to a new location so that you can use the same layout elsewhere. After copying, you will no doubt want to erase the contents of the table so that you can start to fill it out again. An easy way to clear out the table is to select the entire table and then press the **DELETE** key. The information in the table is cleared, but the table structure remains.

It is interesting to note that you must press the **DELETE** key. If, instead, you press **BACKSPACE** while the whole table is selected, Word deletes the entire table, not just the content of the table. (Don't ask me why; it just works that way.)

Entering Tabs in a Table

When you are entering text in a table, some keyboard keys don't function as they do when you are working in the body of your document. One of those keys is the **Tab** key. If you press **Tab**, the insertion point moves to the next cell in the table or, if you are at the end of the table, creates a new row.

At times, however, you may want a tab character in the table's text. To insert a tab, you can use either of these methods:

- Press **Ctrl+Tab**.
- Hold down the **Alt** key as you type 0009 on the numeric keypad.

Keeping Headings in View

If you work with tables, particularly ones that involve many columns, you know that it can be confusing to remember what each column is for. To overcome this problem, divide the current document window into panes. Each pane gives you a view of different parts of your document. In the top pane, display the headings for your columns. You may want to make the pane smaller so that you have more room to work in the other pane. In the bottom pane, do your table work. In this way, you can always see your column headings.

Putting Headings on Your Printout

Most tables have some sort of heading that indicates what is in each column of the table. This heading can be a row or two in size, and you typically apply some sort of special formatting to the heading row (or rows). When your table gets long enough, you may be wondering how you can get your special heading rows to appear at the top of the portion of the table that starts each new page. You do this by following these steps:

1. Select the rows that you are using as your headings.
2. Display the Layout tab of the ribbon.
3. In the Data group, click Repeat Header Rows.

What you see on the screen at this point depends on the view you are using. If you are using Normal view, you will notice no difference in the screen. If you are using Print Layout view, however, the headings appear whenever your table is split across pages.

8

Working with Fields

Word allows you to insert specialized, dynamic information in your documents through the use of fields. Whenever Word sees a field in your document, it examines the contents of the field to derive what is called a field code. The ***field code*** specifies what type of information you want the field to display. Word normally displays the result of the field code in place of the field itself. You can affect how Word displays the information through the use of switches.

You can use literally scores of different fields in your documents. Some of them you will find quite handy; others, esoteric. This chapter will focus on how you use fields within your documents. The first part of this chapter will provide general information about fields, such as how to insert them, display them, select them, and get information about them. The balance of this chapter will present some of the most useful fields provided by Word.

Introducing Fields

Fields can be viewed as nothing but placeholders for other information. You specify the placeholder you want to use, and Word replaces it with the information related to that placeholder. Word provides fields that do everything from providing the user's initials to indicating the filename and size of the document.

In the following sections, you quickly learn how to work with fields. You learn how to insert them, display an actual field or its result, and select a field. You also learn about a handy shortcut that forces Word to change the result of a field to actual text.

Inserting Fields

Word provides a powerful capability to insert special information in your documents through the use of ***fields***. There are scores of fields are available within Word. Inserting a field in your document is very easy; you can use one of two methods:

- Display the Insert tab of the ribbon. In the Text group, click Quick Parts and then choose Field. From the resulting dialog box select the field name from those available. When you click your mouse on OK, the field you requested is inserted in your document.

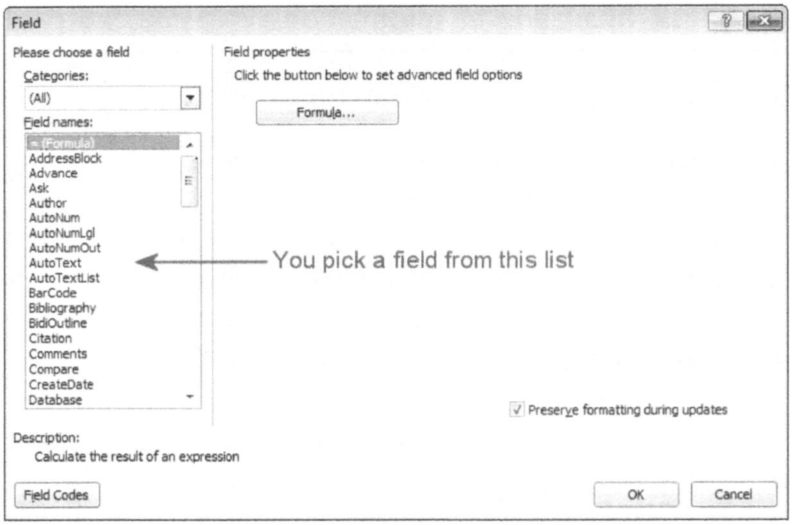

- Press **Ctrl+F9**.

Either of these methods works, but the second inserts only a pair of field braces—you must enter the information and keywords for the actual field. The first method provides a more user-friendly method of inserting fields.

Although these methods work fine for inserting single fields, neither of them will work properly if you want to insert a field within a field. To ***nest*** fields (put one field inside another), you must use the **CTRL+F9** shortcut key to enter the field within the other field.

Displaying Fields

Fields allow the easy inclusion of dynamic data within your documents so they are used quite often in many documents. Normally, Word hides the contents of fields (the "codes" that are intrinsic to the field), replacing them instead with their result, if they are the type of field that yields a result.

If you want to see the contents of your fields, then you can toggle the display so that those contents are visible. You do this by pressing **SHIFT+F9**. Any field contents should then be visible, and you can modify those field contents, if desired.

Remember that this shortcut functions as a toggle. When you are through viewing the field contents, just press **SHIFT+F9** again. Word hides the field contents and replaces them with the field results, if any.

Selecting a Field

When fields are inserted in your document, you might need to select an entire field in order to edit it. The quickest way to do this is to simply select the first character of the field. If field codes are visible, the first character is the opening left bracket. If field codes are not visible (field results are instead visible), you simply select the first character of the result.

When you select the first character, the entire field is selected. This type of selection works whether you are working with displayed field codes or with field results.

Getting Information about Fields

Many of the tips on how to get the most out of Word involve the use of fields. The proper and creative use of fields can increase your productivity with Word quite substantially. The problem is, there is no printed information provided with Word that tells you how to use the fields. (The Word documentation is worthless in this area.) The best way you can get information about fields is to use the on-line documentation in the following manner:

1. Make sure the Insert tab of the ribbon is displayed.
2. Click Quick Parts in the Text group, and in the resulting choices choose Field. Word displays the Field dialog box.
3. Click on the question mark icon in the upper-right corner of the Field dialog box. Word displays its Help system, focused on getting help with fields.

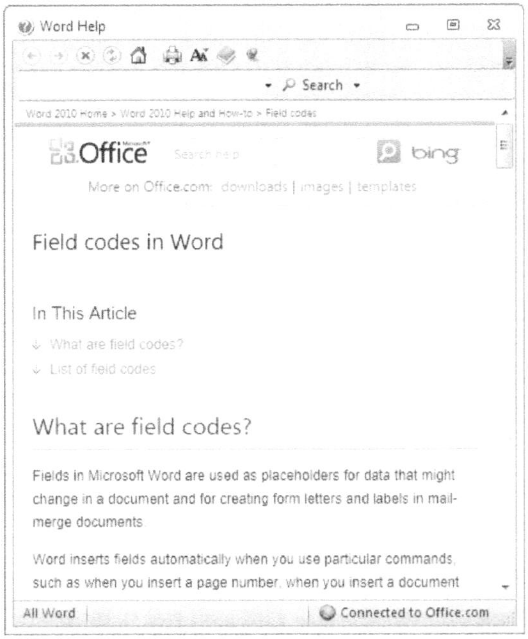

4. Click on List of Field Codes. The Help system displays a comprehensive list of field codes, in alphabetic order.
5. Locate and click on the name of the field about which you want more information.

When you are through reviewing the help information for the field, you can close the Help window and continue your use of Word.

Converting Field Results to Text

Special fields have a great number of uses in Word. You already know that you can use fields to insert everything from the date and time to summary information for a document. (If you don't know this, you will find it out in the next section.) At times, however, you may not want to use a field for information.

Let's say that you have a template that includes a field for today's date in it. When you create a document based on the template, today's date appears in the document. You make changes to the document and then save it to disk.

When you later open the document, it contains, instead of the original date, the date of when you opened the file.

To get around this problem, you need to change the field contents to plain text before you save the file. In this instance, you would follow these steps:

1. Create your new document based on the template. Today's date should appear as normal.
2. Select the date in the document.
3. Press **Ctrl+Shift+F9**. This step converts the field to actual text. Thus, the date is switched from a field to real text for the date.
4. Make any other changes that are necessary in the document.
5. Save the document as normal.

That's it. The next time you open the document, the date remains unchanged. You can also use this technique with any other field to convert it to text.

Inserting Information with Fields

Word 2010 provides scores of different fields (about 75 of them) you can add to your documents. These fields represent a huge amount of potential information that can be added. Not everyone will find all the fields useful. Some fields, however, will be immediately useful to almost everyone.

The next several sections cover some of the more useful fields. You learn how to insert various information, such as the author name, user name, date, current month, important document-related dates, document title and subject, and other statistics.

Inserting the Author's Name

In Word, you can keep track of quite a bit of summary information for a document. One piece you can track is the name of the author of a document. This option is originally set to be the same as the Word user name, but you can change the author name at any time. Word allows you to insert the author's name directly in your document and have it updated automatically if you ever change the author's name. You do this by following these steps:

1. Position the insertion point where you want the author name inserted.
2. Display the Insert tab of the ribbon.
3. Click the Quick Parts tool (in the Text group) and then choose Field. Word displays the Field dialog box.
4. From the Categories list select the Document Information category.

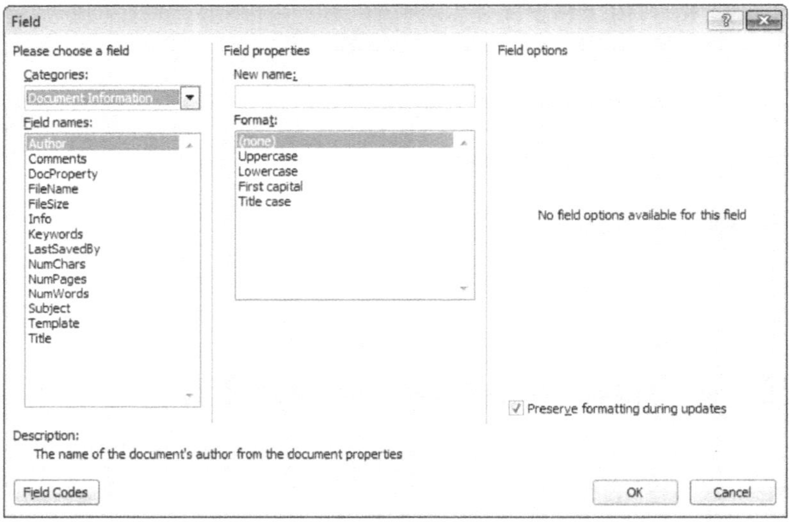

5. Select Author from the Field Names list.
6. Click on OK.

Inserting User Information

When you install Word or Microsoft Office (of which Word is a component) you are asked for your name, along with other information about you. This information is stored with Word and you can later access it. For instance, Word allows you to use your name in the return address of an envelope. You can also insert other information, such as your name or initials. If you want to automatically insert different pieces of user information in your document, follow these steps:

1. Position the insertion point where you want the user name inserted.
2. Display the Insert tab of the ribbon.
3. Click the Quick Parts tool (in the Text group) and then choose Field. Word displays the Field dialog box.
4. In the Categories drop-down list, choose User Information.
5. In the Field Names list, choose UserName to insert the user's name.
6. In the Field Names list, choose UserInitials to insert the user's initials.
7. In the Field Names list, choose UserAddress to insert the user's address.
8. Click on OK.

Inserting Today's Date

Many types of documents require the date. You may be working with a memo, a letter, a report, or some other document in which you must put the date. Word makes this easy by using the Date and Time dialog box. To insert the date, follow these steps:

1. Position the insertion point where you want the date to appear.
2. Display the Insert tab of the ribbon.
3. Click the Date & Time tool, in the Text group. Word displays the Date and Time dialog box.
4. Select a format for the date.
5. Click on OK.

If you choose the "Update Automatically" check box at the bottom of the dialog box, the date is inserted as a field and it will always be updated to reflect the current date as you work on the document on future days. This can be very helpful if you are working with a letter template that you want to always be "in sync" with the day you write the letter.

Inserting the Current Month

In certain documents, it is helpful to insert the current month. For instance, you might need to do a monthly report in which you place the current month name in the title of the report. Word allows you to insert a field for the month, and this field is updated automatically whenever you print your document. To do this, follow these steps:

1. Position the insertion point where you want the month inserted.
2. Press **Ctrl+F9** to insert a set of field braces. Make sure the insertion point remains between the braces.
3. Type **date \@MMMM**.
4. Press **F9** to update the field.

Inserting the Document Creation Date

Word keeps track of a good deal of information about your document and then makes that information available to you. One piece of information that is tracked is the document creation date. It is the date when the file was first created (opened as a new file) or the last time you chose Save As and saved the file under a new name. Word allows you to insert this date into your document by following these steps:

1. Position the insertion point where you want the date to appear.
2. Display the Insert tab of the ribbon.
3. Click the Quick Parts tool (in the Text group) and then choose Field. Word displays the Field dialog box.
4. From the categories of fields, choose Date and Time.
5. Select CreateDate from the Field Names list.
6. Select a date format by clicking on the Options button, if desired.
7. Click on OK to insert the field.

This type of date field is a big help when you are using dates with memos, letters, and reports. The date stays the same as when you first created the file, unless (again) you save the file under a new name, which changes the file creation date.

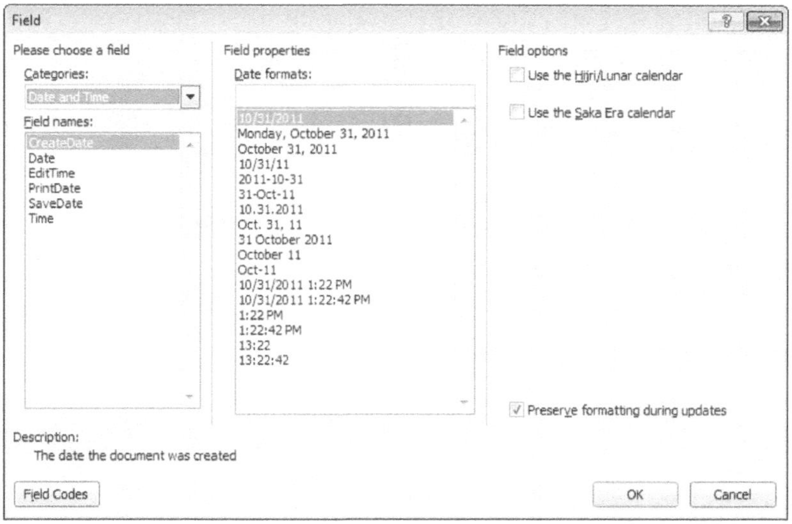

Inserting the Date Your Document Was Last Saved

Every time you save a document in Word, some of the document properties information is automatically updated. Part of the information is the date on which the document was last saved. You can insert this date into your documents and have it updated automatically, if you want. To insert the last date your document was saved, follow these steps:

1. Position the insertion point where you want the date to appear.
2. Display the Insert tab of the ribbon.
3. Click Quick Parts (in the Text group) and then click Fields. Word displays the Field dialog box.
4. Choose Date and Time from the field categories at the left of the dialog box.
5. Select SaveDate from the field list at the right side of the dialog box.
6. Click on Options to select a date format, if desired.
7. Click on OK to insert the field.

Inserting the Date Your Document Was Last Printed

One piece of information that Word maintains about documents is when they were last printed. Every time you print your document, this piece of information is updated. Sometimes, you may want to display this "last printed" date in your document. To do this, follow these steps:

1. Position the insertion point where you want the date to appear.
2. Display the Insert tab of the ribbon.
3. Click Quick Parts, in the Text group, and then choose Field. Word displays the Field dialog box.
4. Choose Date and Time from the Categories list. Word updates the choices in the Field Names list.
5. Select PrintDate from the field list at the right side of the dialog box.
6. Click on Options or Field Codes to select a date format, if desired.
7. Click on OK to insert the field.

Inserting the Edit Time

Word keeps track of how long you work with a file. You can see this information on the Statistics tab of the Properties dialog box. As far as Word is concerned, the cumulative amount of time you have the file open is considered the editing time for the document. You can insert this time, in minutes, in your document by using the following steps:

1. Position the insertion point where you want the editing time inserted.
2. Display the Insert tab of the ribbon.
3. Click the Quick Parts tool (in the Text group) and then click Fields. Word displays the Field dialog box.
4. Choose Date and Time from the Categories list.
5. Select EditTime from the list of available fields.
6. Click on OK.

Inserting the Document Title in Your Document

Word keeps track of many pieces of information about your document. You can see the information maintained by Word simply by displaying the File tab of the ribbon and then clicking Info | Properties | Advanced Properties.

If you search around through the information that Word maintains about a document, you'll find that Word provides the opportunity to specify a title for your document. Once you set the title, Word allows you to insert this information into your document and have it updated automatically. This means you can set the title once in the document properties, and then have it inserted in the header, footer, or heading for your document. That way, you only have to update it in one place and it will automatically be updated everywhere in the document.

To insert the document title into your document, follow these steps:

1. Position the insertion point where you want the title to appear.
2. Display the Insert tab of the ribbon.
3. Click the Quick Parts tool in the Text group and then choose Field. Word displays the Field dialog box.
4. In the Categories list, choose Document Information. Word updates the choices in the Field Names list (right side of the dialog box).
5. In the Field Names list choose Title.
6. Click on OK to close the dialog box and insert your field.

Inserting the Subject in Your Document

Word keeps track of properties for a document. You can see the information maintained by Word by displaying the File tab of the ribbon and then clicking Info | Properties | Advanced Properties.

One of the pieces of information you have the opportunity to enter in these properties is the subject of your document. Once you set the subject, Word allows you to insert this information into your document by following these steps:

1. Position the insertion point where you want the subject to appear.
2. Display the Insert tab of the ribbon.
3. Click the Quick Parts tool in the Text group and then choose Field. Word displays the Field dialog box.

4. In the Categories list, choose Document Information. Word updates the choices in the Field Names list (right side of the dialog box).
5. In the Field Names list choose Subject.
6. Click on OK to close the dialog box and insert your field.

Inserting a Page Number Field

Word provides a few shortcuts that allow you to insert common fields into your documents. If you have a need to insert the PAGE field often, then you will be interested in this shortcut. (The PAGE field automatically reflects the page number on which it is placed. For instance, if the PAGE field is inserted on page 4 of your document, it is replaced with a 4. If editing subsequently changes that page on which the PAGE field is located, the field is replaced with the then-current page number when your fields are updated.)

To insert a PAGE field where your insertion point is located, simply press **Shift+Alt+P**. The PAGE field is inserted, updated, and automatically formatted with the Page Number character format.

You should note that it is possible for the **Shift+Alt+P** shortcut to be "hijacked" by some other part of your system. For instance, in some versions of Windows pressing this shortcut key brings up the Windows Media Player; Word never gets a chance to act upon the shortcut. In that case, there is nothing you can do (short of removing, deactivating, or reprogramming whatever program is hijacking the shortcut key) except to insert the PAGE field the old fashioned way—by the ribbon tools to insert the field.

Inserting Document Statistics

Word keeps track of summary information for a document. You can see the information maintained by Word by displaying the File tab of the ribbon and then clicking Info | Properties | Advanced Properties. Part of the information maintained by Word is the number of words in the document, as well as the number of characters and the document size. These tidbits are updated whenever the document is saved or printed. Word allows you to insert these items in your document by following these steps:

1. Position the insertion point where you want the number of words to appear.
2. Display the Insert tab of the ribbon.
3. Click the Quick Parts tool in the Text group and then choose Field. Word displays the Field dialog box.
4. Using the Categories drop-down list, choose Document Information.
5. To insert the number of words, select NumWords from the Field Names list.
6. To insert the number of characters, select NumChars from the Field Names list.
7. To insert the file size, select FileSize from the Field Names list.
8. Click on OK to close the dialog box and insert the field.

You should note that the document's file size (see step 7) is different from the number of characters in the document (see step 6). The file size is invariably larger because it includes space consumed by formatting and housekeeping information maintained by Word.

Numbering Elements Sequentially in Your Document

Word allows you to automatically number items in your document. For instance, you can automatically number figures, tables, or illustrations. If you later reorder these items, Word automatically updates the numbers so that they are in order. To number items sequentially in your text, follow these steps:

1. Position the insertion point where you want the sequential number to appear. For instance, this would be in the caption for the table or figure.
2. Press **Ctrl+F9** to insert field brackets. Make sure the insertion point stays between the brackets.
3. Type "**seq** **"** followed by the name of the element. This name is up to you, but should be the same for each item in this sequence. For instance, you could type "**seq figures"** or "**seq tables**" (without the quote marks).

4. Press **F9** to update the field information. Word replaces the field with the next number in the sequence you have specified.

9

Printing Documents

The ultimate goal of most people who create documents in Word is to put together something that looks great when it's printed. For this reason, Word spends a great deal of its processing time making sure that what you see on your screen matches what you print on your printer.

In this chapter, you learn about how to make Word work with your printer. For the most part, Word uses any printer you successfully install under Windows. However, you can focus on other things. For instance, you can learn how to control your printer from within Word. In this chapter, you learn about performing print jobs, setting up the printer, printing only portions of a document, and taking advantage of special printer features.

Controlling Your Printer

Your printer is an integral part of your computer system. If you want to print a Word document to your printer, you can typically do so by simply pressing **Ctrl+P** or by displaying the File tab of the ribbon and clicking Print. Word, however, allows you to control exactly how it interacts with your printer after you take either of those actions.

In the following sections, you learn all about how to print in the background, adjust your printer's properties, print on both sides of the paper, print a draft of your document, and cancel printing.

Understanding Background Printing

Word includes a feature called ***background printing.*** When you use this feature, Word spools your document to disk and then feeds it to the Windows printing system as fast as possible. While the document is being printed, you are free to do other tasks within Word. When background printing is turned off, Word stops everything and sends your document to Windows right away.

If you are working with short documents, you will notice little difference in using background printing. If you are printing long documents, however, you may see quite a bit of difference. With background printing turned on, you can continue to work in Word even when the program is sending a long document to the printer. This feature can be a huge benefit. On the other hand, if you have background printing turned off, your long documents may print faster, but you cannot use Word while the printing is being done.

To control background printing, follow these steps:

1. Display the File tab of the ribbon and then click Options. Word displays the Word Options dialog box.
2. At the left side of the dialog box click Advanced.
3. Scroll through the options until you see the Print section.

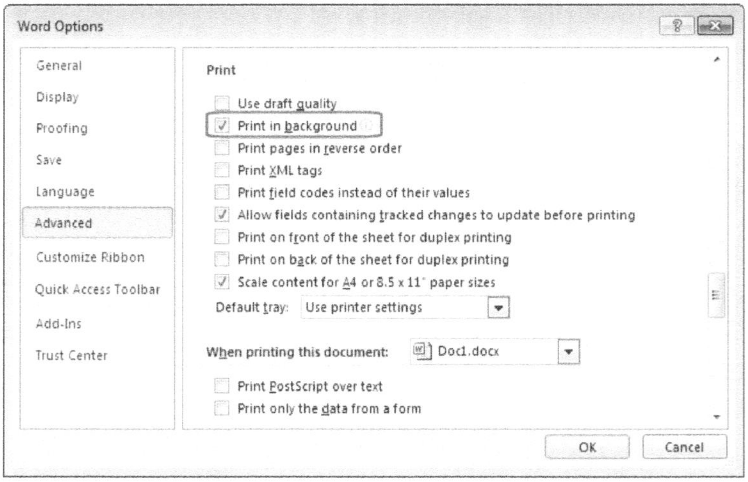

3. Note the setting of the Print in Background check box. If it is selected, background printing is enabled; if it is cleared, background printing is turned off.
4. Click on OK when you are finished.

Setting Up Your Printer

To change the settings Word uses when printing through your printer, follow these steps:

1. Press **CTRL+P**. Word displays the File tab of the ribbon with the print options visible.
2. Using the Printer drop-down list, select the printer you want to use. (It is common to have more than one printer available if you are working in a networked environment.)
3. Click on the Printer Properties link. Word displays the Properties dialog box for the printer you selected in step 2. The exact appearance of the dialog box, and the controls it contains, depends on the printer driver you have installed for your printer.

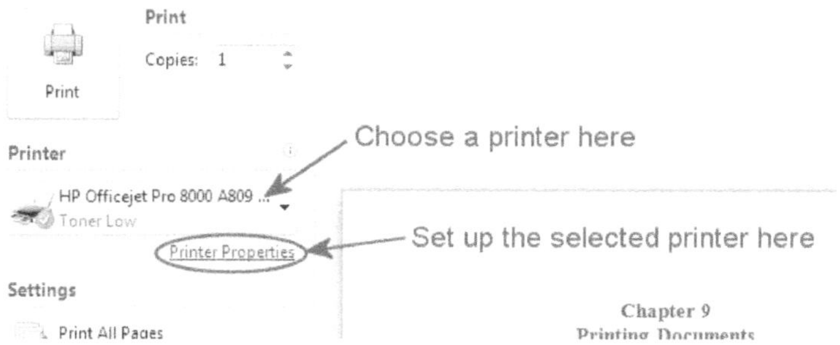

4. Using the tabs and other controls in the dialog box, specify the printer settings you want to use.
5. Click on OK to close the Properties dialog box.
6. Click on Print. Your document is printed.

Using Duplex Printing

Some printers support ***duplexing,*** meaning that they allow printing on both sides of a piece of paper. Being able to print on both sides depends on what type of printer you have set as your default printer. For instance, an HP OfficeJet Pro L7780 prints duplex but other models of HP printers may not offer this as an option.

Assuming that you have a printer that supports duplexing and that you have a printer driver that supports it, you can turn it on by following these steps:

1. Press **Ctrl+P**. Word displays the File tab of the ribbon with the print options visible.
2. Using the Printer drop-down list, select the printer on which you want the document printed.
3. Click on the Printer Properties button. Word displays a Properties dialog box for the printer you selected in step 2.
4. Select the Paper tab, then click the More Options button.
5. Change the Duplex Printing option to either Flip on Long Edge or Flip on Short Edge, depending on how you want your document printed.
6. Close the dialog boxes and print your document.

Again, these steps vary based on the printer you are using. You may need to look around in the printer's Properties dialog box to see where the actual control for duplexing is located.

Printing a Draft of a Document

Word allows you to print drafts of your documents. This feature can be very helpful in speeding up the printing process, particularly if you have a complex document with lots of fonts and graphics or a slow printer (or both). When printing drafts, the final fonts are not used, nor are the graphics printed in place. To print a draft, follow these steps:

1. Display the File tab of the ribbon and then click Options. Word displays the Word Options dialog box.
2. At the left side of the dialog box click Advanced.

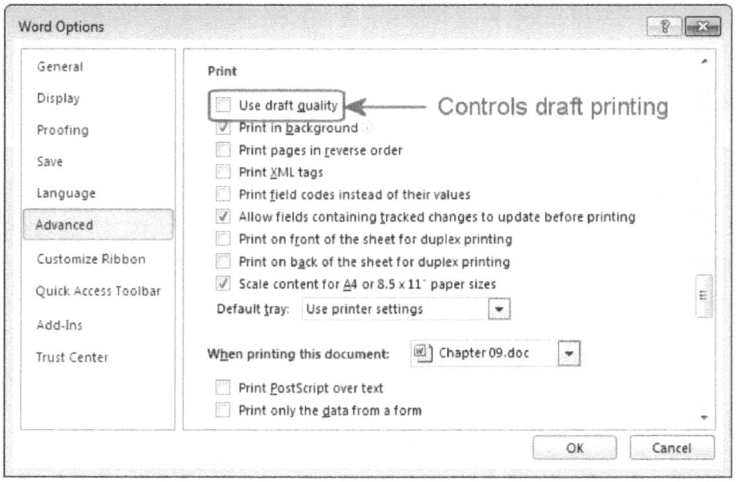

3. Scroll through the available options until you see the Print section.
4. Make sure the Use Draft Quality check box is selected.
5. Click on OK.
6. Print your document.

Remember that the exact way in which draft printing is implemented is, in large part, up to your printer. Don't be surprised if you get different results doing a draft print on one printer as compared to another.

Canceling Printing

If you are printing a long document, you may decide that you want to cancel printing for some reason. Exactly how you do this depends on if you are using the background printing feature of Word. If you have background printing turned off, Word displays a dialog box during the printing process. This dialog box, besides informing you of the progress of the print job, also contains a Cancel button. Simply click on this button to stop the print job.

Most people, however, will have background printing turned on—this is the default setting for Word. With background printing turned on Word does not display a dialog box. However, on the status bar you can see a small printer icon that indicates the page currently being printed. (If you have a fast printer and a relatively short document, then the icon may not appear long enough to see.) To cancel the print job, double-click on the printer icon.

You should remember that print jobs can be stored at various places before they are actually printed. Word "hands off" the print job to Windows, and then Windows figures out what to do with it. Most likely it sends the job to the printer (if the printer is directly connected to your computer) or it sends it to another location on your network, such as another computer or the printer itself. Why is this important? Because if you want to cancel a print job, you may not be able to do it directly from Word—you may need to figure out where the print job is located on its travels to the printer and then cancel it at that location.

Setting Up a Print Job

A ***print job*** is technospeak for something being printed—and it sounds better than saying "Setting Up Something Being Printed." Print jobs are created when you tell Word to print something. In the following sections, you learn how to tell Word exactly what you want to print and how you want it printed. In other words, you learn what settings you can specify before you click on the Print button.

Previewing Before You Print

Word has a useful feature that allows you to see what your printed output will look like before you print. This feature, called ***print previewing***, is automatically done by Word when you get ready to print. In other words, when you press **CTRL+P** and Word displays the File tab of the ribbon, a print preview of your document is displayed.

You can move around the print preview by clicking the page number controls just under the preview. The preview of your document differs from the other viewing modes of Word in that it shows an exact representation of what your printed output will look like. The other viewing modes only provide different levels of approximation.

When examining your preview, you can view your document closer by using the zoom slider at the bottom-right corner of the preview. Slide it left or right, depending on whether you want to shrink or expand the preview.

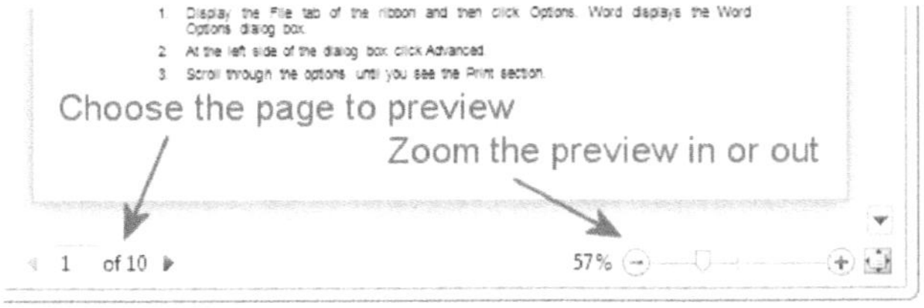

When you are done viewing your document in print preview mode and you decide that you don't want to print it, you can exit by using either of these methods:

- Press Esc.
- Click one of the other tabs on the ribbon, such as the Home tab.

Collating Copies

If you work at a place that has a good-size copy machine, you already know what it means to collate something. With a copier, ***collate*** means that the document you are copying is placed in complete sets that can be used right away. It is the same in Word. If you have a document that occupies five pages and you are printing multiple copies, you can print them either collated or normally. When they are collated, they are in page 1–5 order for each of the five sets. If they are not collated, five copies of the first page are printed, and then five of the second, and so on.

To control collating in Word, follow these steps:

1. Press **CTRL+P**. Word displays the File tab of the ribbon with the print options visible.
2. Click on the collation control, under the Settings area of the page.
3. Choose which collation method you want to use—collated or uncollated.
4. Click on Print. Your document is printed.

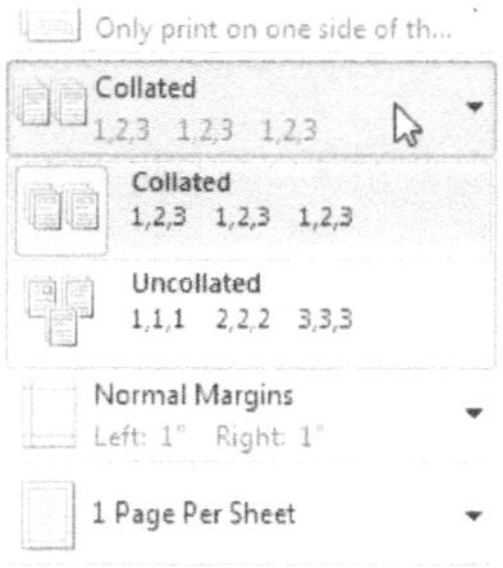

Printing Only Selected Pages

Word gives you complete control over how it prints your document. You can print either the entire document or just a portion of it. To print only selected pages, follow these steps:

1. Press **CTRL+P**. Word displays the File tab of the ribbon with the print options visible.
2. In the Pages box, indicate the pages you want to print.
3. Click on OK.

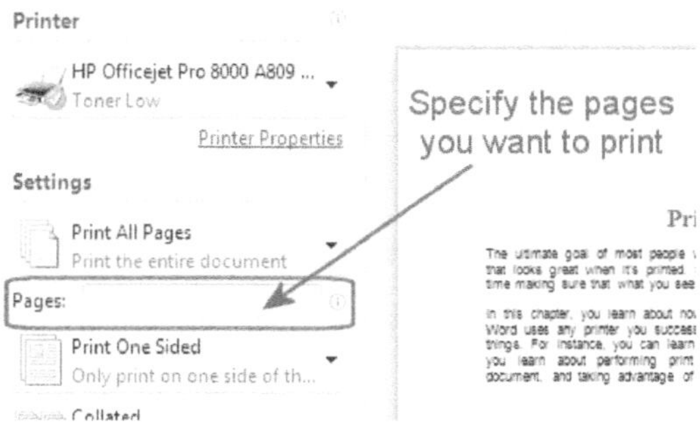

When you enter the number of pages in step 2, you can enter them in a variety of ways. The following table illustrates the different ways you can specify which pages you want to print:

To Print This	Enter This in the Pages Box
A range of pages	Type the beginning and ending page numbers with a hyphen in between. For instance, **3-7** prints pages 3 through 7.
Noncontiguous pages	Type the page numbers separated by commas. For instance, to print pages 2, 5, and 12, you enter **2, 5, 12**. To print a range as part of what is printed, you can do so in the manner of **2, 5-8, 12**.
A range of pages in a specific section	Use **p** and **s** to indicate page and section. For example, **p3s2-p7s2** prints pages 3 through 7 in section 2.
An entire section	Use **s** followed by the section number. For example, **s2** prints all of section 2.
Noncontiguous sections	Use commas between the sections you want to print. For example, **s2, s4, s6** prints all of sections 2, 4, and 6.
A range of pages across sections	Type the beginning page and section and the ending page and section separated by a hyphen. For instance, **p3s2-p7s4** prints everything between page 3 of section 2 and page 7 of section 4.

Printing More Than One Copy

Word gives you complete control over how it prints your document. If you want, you can print more than one copy of your document. To print to a file, follow these steps:

1. Press **CTRL+P**. Word displays the File tab of the ribbon with the print options visible.
2. In the Copies box, indicate the number of copies you want to print.
3. In the Range box, specify what you want to print. (See the preceding section.)
4. Click on OK.

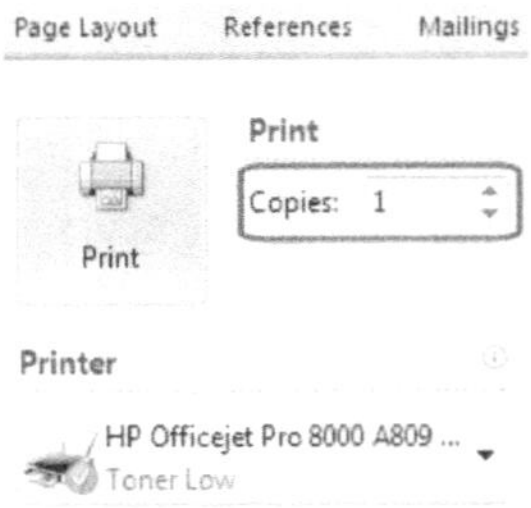

Printing Odd or Even Pages

Many times, you may need to print either odd or even pages in a print job. For instance, you may want to put your pages through the printer twice so that you can print on both sides. This task is easy to do in Word by following these steps:

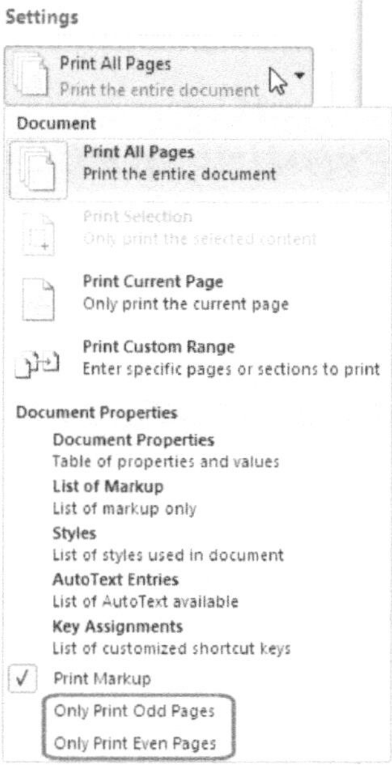

1. Press **Ctrl+P**. Word displays the File tab of the ribbon with the print options visible.
2. Adjust the printing settings as desired.
3. Click the first button under the Settings heading and Word displays a series of options that control what it prints.
4. Select either Only Print Even Pages or Only Print Odd pages, as desired.
5. Click on Print. Your document is printed.

Reversing the Print Order

When printing your document, Word allows you to print it either forward or backward. This option is helpful if you have a printer that deposits printed pages one on top of the other, right side up, as they are printed. This arrangement is typical with certain models of inkjet and laser printers. When the printing order is reversed, the final document is collated properly. To control the printing order, follow these steps:

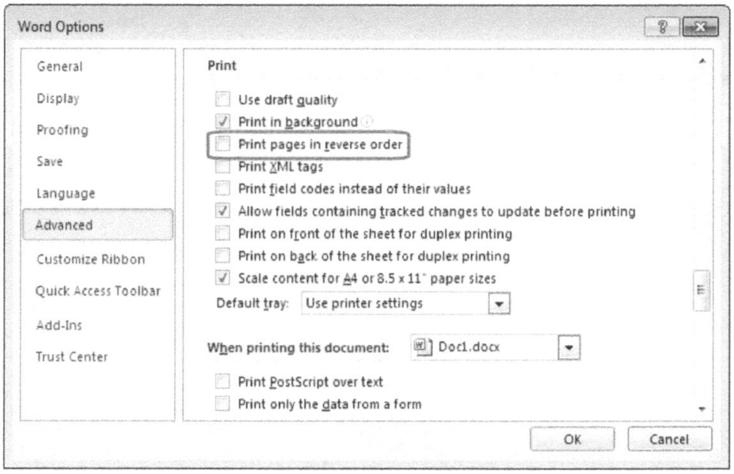

1. Display the File tab of the ribbon and then click Options. Word displays the Word Options dialog box.

2. At the left side of the dialog box click Advanced.
3. Scroll through the options until you see the Print section.
4. Make sure the Print Pages in Reverse Order check box is selected or cleared, depending on your desires.
5. Click on OK.

Printing Multiple Pages per Sheet

Word has a feature that allows you to print multiple pages on a single sheet of paper. For instance, you can print two pages of your document on a single side of a piece of paper. Word allows you to print 1, 2, 4, 6, 8, or 16 actual pages per printed page. This option can save quite a bit of paper, but it can make reading the output quite a bit more difficult. To control this setting, follow these steps:

1. Load the document you want to print.
2. Press **Ctrl+P**. Word displays the File tab of the ribbon with the print options visible.
3. Use the Pages Per Sheet drop-down list, at the bottom of the settings, to specify how many pages should be combined on each printed page.
4. Specify any other printing options, as desired.
5. Click on OK to print your document.

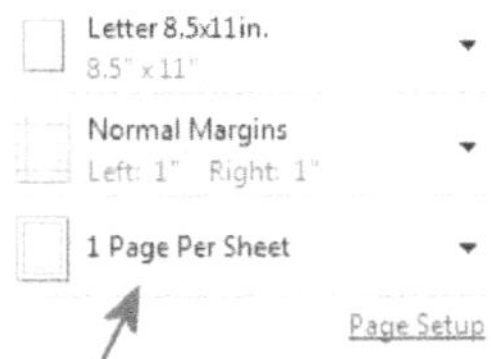

Controls the number of pages printed per single sheet of paper

Getting a Double-Spaced Printout

It is not uncommon to need a double-spaced printout of a document while you are developing it. However, it can be bothersome to save the document as double-spaced, particularly if the need for double spacing is only temporary (such as when your final document will be single-spaced). If you need an easy way to print a double-spaced document, follow these quick steps:

1. Save your document.
2. Press **Ctrl+A**. Your entire document is selected.
3. Press **Ctrl+2**. Word double-spaces the document.

4. Print your document as you normally would.
5. Close your document without saving

This last step is important. If you save your document before exiting, the double spacing is permanently saved with the document as well.

Printing to a File

Word gives you complete control over how it prints your document. You can print to either a printer or a file. To print to a file, follow these steps:

1. Press **CTRL+P**. Word displays the File tab of the ribbon with the print options visible.
2. Click the Printer drop-down list (the one you normally use to specify the printer you want to use) and at the bottom of the list, choose Print to File.
3. Set other printer options, as desired.
4. Click on Print. Word asks you to provide a filename for the output.
5. Enter a filename, including a path, if desired.
6. Click on OK.

After the output has been sent to a file, you (or someone else) can later copy the file to the destination printer from the Windows command line by using the copy command. Assume that you have created an output file named Report.Prn. You could later send this file to the printer connected to your parallel port by using the following command at the command prompt:

```
copy report.prn lpt1:
```

Special Printing Information

To get the best possible output, you must be concerned with more than just how your document appears in Word. You must also be concerned with such mundane matters as the type of paper on which you will print and what size labels you want to use. In the following sections, you learn how to approach those issues as well as how to prepare a document to be printed by a service bureau.

Preparing Files for a Commercial Printer

When it comes time to do your final output, you may want to consider working with a service bureau to prepare camera-ready copy. A ***service bureau*** is a company that will take your files and output them on a high-resolution printer or phototypesetting system. *Camera-ready copy* is final output that is suitable for final printing. A service bureau could be anyone, ranging from your local print shop to a national commercial printer.

If you decide to work with a service bureau, make sure you contact them as early as possible. You will need to discuss the following points with them to make sure you both understand what is necessary in order to get the highest quality final product:

- What fonts are used in your document? Does the service bureau have the same fonts, from the same font vendor?
- What printer driver are you using? Does it match the printer driver used by the service bureau to output to its equipment?
- What is the size of your final printed output? Will the bureau's equipment handle the dimensions you are using?
- Exactly what does the service bureau need from you? Does it need the original Word files, a PDF created from your Word files, or perhaps a PostScript output file?
- When must you have the files to the service bureau in order to receive the camera ready copy when you need it?

When looking for a service bureau, consult the Internet or the Yellow Pages under the heading *typesetting* or *typesetters*.

Following Guidelines for Laser Printer Letterhead

Because of the way laser printers place ink on paper, you must make some special considerations when you're ordering letterhead for use in the printer.

If you don't follow these guidelines, you can ruin a good batch of paper or, worse, ruin the laser printer itself.

First, if you like your letterhead printed using thermography, you should realize that thermography and laser printers don't mix. (***Thermography*** is a printing technology that allows the ink to rest on the surface of the paper. This process is often used for business cards and letterhead.) Thermography is applied at a relatively low temperature and therefore has little heat tolerance. The high heat used in a laser printer causes the ink to come off the paper and adhere to the laser printer's drum. This situation damages the drum and affects all subsequent print quality.

Some people also like to use metallic foils in their letterhead. These foils catch the light very nicely and provide a "rich" appearance. Foil is also applied with heat when printing the letterhead, but usually it is under a higher heat than what is used in thermography. Even so, few foils will hold up under use in a laser printer. It takes an experienced foil stamper to accomplish a successful foil application to stationery for the purpose of laser printing. The first item required for this purpose is the proper-weight material—nothing less than a 24-pound stock should be used. Anything lighter cannot absorb enough heat from your laser printer to keep the foil from flaking off when passing the laser drum.

The type of paper stock used with the foil is also important. The paper should be a bond, cotton, rag, or other good material. The most critical element, however, is the type of foil that is used. In fact, without the correct foil, no other step matters.

As far as using embossed letterhead, the same general considerations apply. The heavier the material, the better, and the type of embossing and the type of die that are used are very important. For example, a beveled embossing works better than a domed embossing, and copper or brass is much better than magnesium. Heat is also very important while embossing in order to hold a good embossing when printed on a laser printer.

If you are considering new letterhead, you should always talk to your printer and see whether he or she can provide samples of similar letterhead that you can test before you spend the money to print your order.

Using Good-Quality Labels

If you print labels in your printer, make sure to purchase labels that are of good quality and that are designed for the printer in which they will be used. If you have a dot matrix printer, low-quality or marginal labels may peel off as the label stock moves around the platen. If this happens, you could end up with labels stuck inside your printer, where it is difficult to remove them.

If you have a laser printer, make sure the labels you use are designed for the high heat generated by a laser printer. (The label packaging should indicate that the labels are designed for laser printers.) If not, the adhesive on the back of the labels may soften to a point where the labels come off in the printer or the adhesive oozes out and messes up your printer. Good-quality labels should be available from any business supply store and are worth the few extra pennies you have to pay.

Adding a Watermark

A ***watermark*** is normally light printing that appears behind your normal text, although it could also be a light version of a graphic. Watermarks are great for adding words such as "Draft" or phrases such as "Do Not Copy" to each page of your printout.

Word 2010 allows you to allows you to easily add watermarks to your document. Follow these steps:

1. Open the document to which you want to add a watermark.
2. Display the Page Layout tab of the ribbon.
3. Click the Watermark tool in the Page Background group. Word displays an assortment of watermarks from which you can choose.
4. Click the watermark you want added to your document.

That's it; the watermark is added and you should see it appear behind the text of your document. If you don't like any of the watermarks available, you could click the Custom Watermark option and define your own. (Perhaps you want to create one that includes your company name.)

If, at some point, you want to remove the watermark, just click the Watermark tool and then choose Remove Watermark.

Printing Field Codes

If you use fields in your documents to any extent, you may want to print a copy of your document with field codes showing instead of field results. (Field codes are discussed in Chapter 8, "Working with Fields.") To print field codes instead of results, follow these steps:

1. Display the File tab of the ribbon and then click Options. Word displays the Word Options dialog box.
2. At the left side of the dialog box click Advanced.

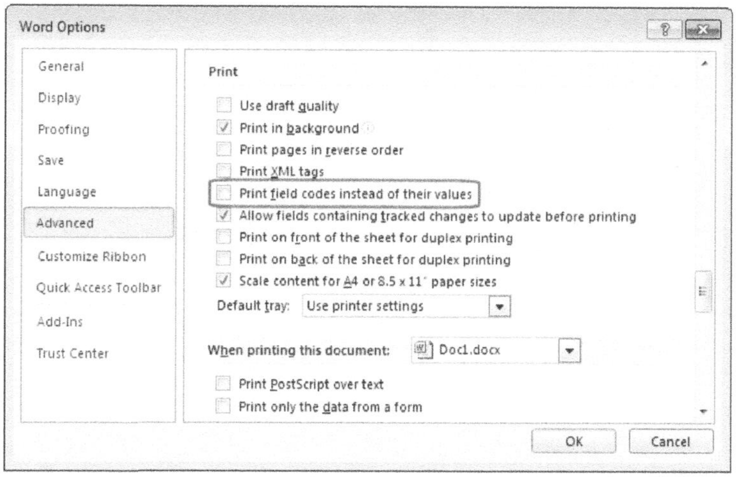

3. Scroll through the options until you see the Print section.
4. Make sure the Print Field Codes Instead of Their Values check box is selected.
5. Click on OK.
6. Print your document.

10

Managing Documents and Files

As you use Word for a while, you begin to notice a curious thing: You start accumulating files—lots of files. Old copies of reports, letters, and forms begin to clutter your hard drive. It is not unusual to accumulate multiple versions of the same document, each slightly different from the other.

Along with this accumulation of files comes the need to manage those files. Some information related to file management is considered under the realm of Windows, not Word. For instance, creating folders, renaming files, copying files, and deleting files are typically done in Windows, not in Word.

You can do different things, however, within Word itself to aid you in your management efforts. The purpose of this chapter is to share some of that information with you. Here, you learn how to quickly load files you have recently worked with, open multiple documents at the same time, and open documents as read-only. You also learn that you can control how Word does its file saving.

Understanding Recent Documents

Most people who use Word find that they work regularly with a limited number of documents. Whereas you might work with hundreds of documents over the course of a year or two, in the short-term you probably only work with ten or fifteen different documents.

Fortunately, Word keeps track of the documents on which you work. This can come in very handy because it allows you to quickly access the documents on which you most recently worked. To access this list of documents, display the File tab of the ribbon and click on Recent at the left side of the screen.

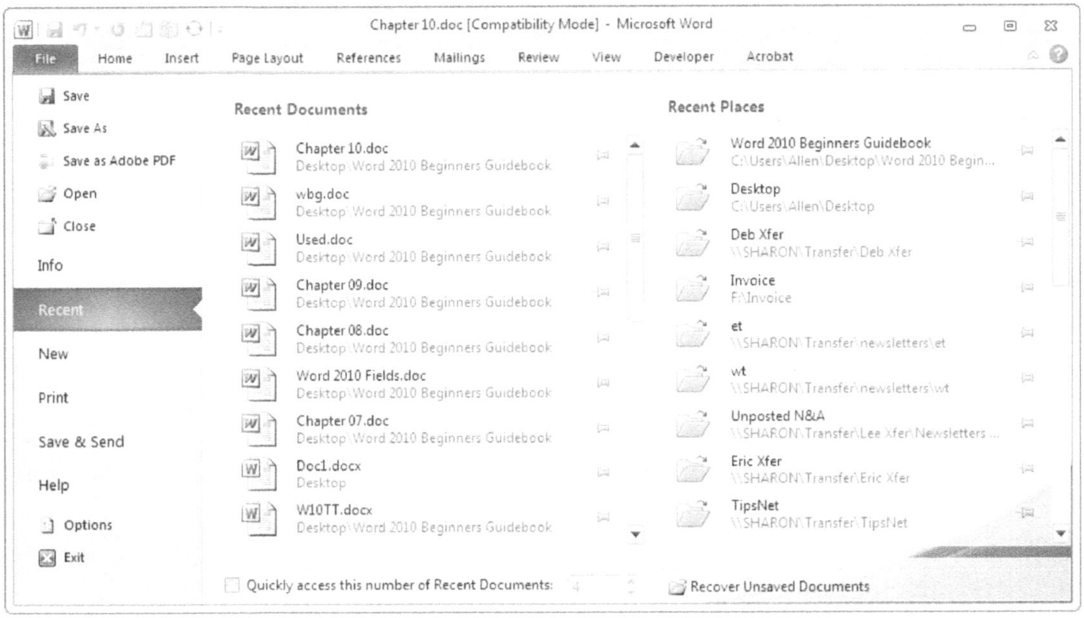

To open a document, simply click on it in the list of recent documents. Word opens the document and allows you to edit it like normal.

There is something on the recent documents screen that you should note. If you look just to the right of each document in the list, you see an icon that looks like a thumbtack or a push pin. If you click this icon, then the file is "pinned" to the top of the recent documents list. Remember that as you use documents, the list of recent documents is constantly in flux—new documents appear at the top of the list and documents you haven't used in a while drop off the bottom of the list.

Pinning a document is a great feature if you want to always have a file (or number of files) accessible through the list, when they otherwise may fall off the bottom of the list. This ensures that you can get to your most active documents at any time you need.

Adjusting the Number of Recent Documents

You can easily instruct Word to maintain more (or less) documents in your recent documents list. You do that by following these steps:

1. Display the File tab of the ribbon and then click Options. Word displays the Word Options dialog box.
2. At the left side of the dialog box click Advanced.
3. Scroll through the options until you see the Display section.

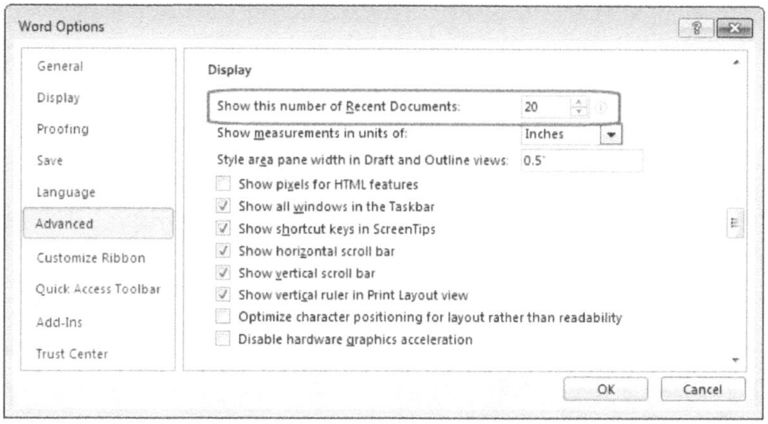

4. Adjust the Show This Number of Recent Documents setting to reflect how many you want Word to keep.
5. Click OK.

In step 4 you can specify a value from 0 to 50. If you start out with a value (let's say 20), lower the value to a different value (let's say 5), then Word only starts tracking the most recent five files. If you subsequently raise the value again (let's say back up to 20), then Word doesn't immediately raise the number of documents in the recent documents list. It only raises the number over time, as you access additional documents, until the list is filled out at twenty.

Displaying the MRU List

MRU is an acronym for ***most recently used.*** It describes a feature of Word that allows you to quickly open your most recently used documents. Don't confuse this feature with the recent documents list; the MRU list is best viewed as a subset of the recent documents. The MRU list is displayed at the left side of the File tab of the ribbon, just the Close command and above the Info command.

To display the MRU list or to change the number of documents shown in the MRU list, follow these steps:

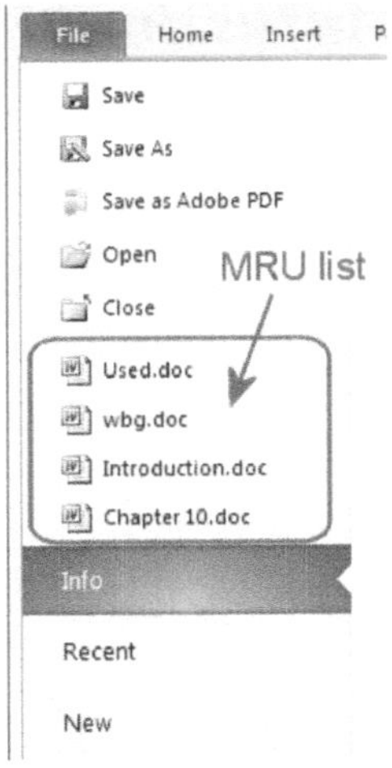

1. Display the File tab of the ribbon and then click Options. Word displays the Word Options dialog box.
2. At the left side of the dialog box click Recent.
3. At the bottom of the screen, click the check box labeled Quickly Access This Number of Recent Documents. You should see the MRU list immediately appear at the left of the screen.
4. Adjust the value shown to the right of the check box, if desired.

Opening Multiple Documents at a Time

Word provides a complete file management system as part of Word. One feature of this system is that you can quickly load multiple documents all at one time. To do this, follow these steps:

1. Display the File tab of the ribbon and then click Open. Word displays the Open dialog box.
2. Instead of double-clicking on a filename, start to create a "selection set." If you are selecting sequential files, click on the first file in the sequence and then hold the **Shift** key as you click on the last. You can add additional files to those selected by holding down the **Ctrl** key while clicking with the mouse.
3. To remove files from the selection set, hold down the **Ctrl** key while clicking on a filename you previously added to the set.
4. When you are satisfied with the list, click on Open.

Opening a Document As Read-Only

Sometimes, you may want to open a file as ***read-only*** so that you cannot save any changes to the file you are loading. You can use the Save As option from the File tab of the ribbon to save the document to a different filename, however. Opening a file as read-only is self-insurance so that you don't inadvertently mess up your file. To open a document as read-only, follow these steps:

1. Display the File tab of the ribbon and then click Open. Word displays the Open dialog box.

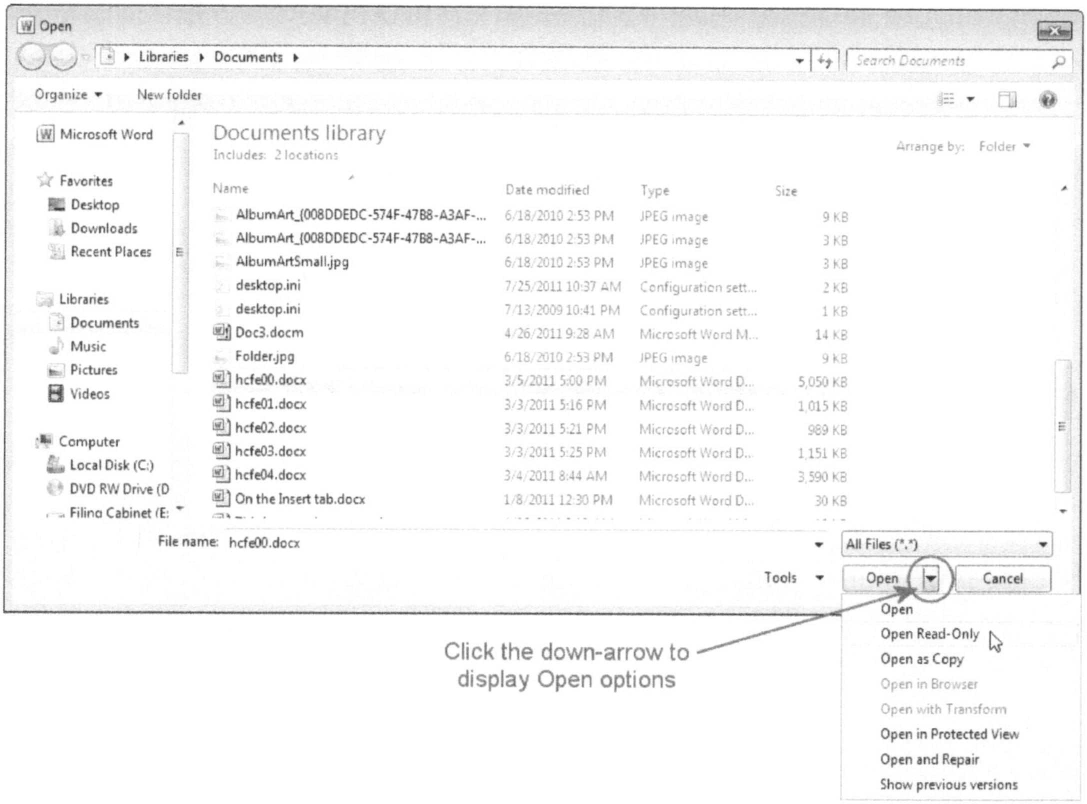

2. Locate the file you want to open, and then click once on the filename.
3. Click on the down-arrow at the right of the Open button. Word displays a drop-down list showing you different ways you can open your document.
4. Choose Open Read-Only from the drop-down list.

Quickly Managing Files

Many times, you need to do a little rudimentary file management from within Word. For instance, you may be through working with a file and need to delete it. You can perform such rudimentary tasks very easily by following these steps:

1. Display the File tab of the ribbon and then click Open. Word displays the Open dialog box.
2. Use the tools in the dialog box to display the folder that contains the files you want to manage.
3. To delete a file, select it and press the DELETE key. You can also right-click on the filename and choose Delete from the Context menu.
4. To rename a file, right-click on it and choose Rename from the Context menu.
5. To copy a file, right-click on it and choose Copy from the Context menu. Then right-click in an empty area of the file list, and choose Paste from the Context menu.

Saving Your Work Automatically

We've all done it: You're working along for several hours on a document, when suddenly the power goes out. Or you kick the reset button on your computer, or your three-year-old pulls the power cord out of the wall. The list goes on, but the bottom line is that you have lost the past several hours of work. Many things spring to mind at a time like this, and most of them can't be printed in a nice, family-oriented book such as this one. But, hopefully, you learned a lesson when this situation happened to you.

Word allows you to protect yourself by automatically saving your work for you. To set the automatic saving feature, follow these steps:

1. Display the File tab of the ribbon and then click Options. Word displays the Word Options dialog box.
2. At the left side of the dialog box click Save.

3. Click on the Save AutoRecover Info Every check box. The number 10 appears in the Minutes box.
4. Adjust the Minutes box to reflect how often you want Word to save your document. You can select values between 1 minute and 120 minutes.
5. Click on OK.

When designating an AutoRecover frequency, you should probably not select a time under 10 minutes. More frequent saves can waste time and become counterproductive.

You should also note that AutoRecover does not really save your file. What it does is to save information that Word can use to try to recover your file if Word stops unexpectedly. If Word detects AutoRecover information when it first starts (such as after the power went out), it asks whether you want to recover it or discard it.

Understanding Background Saving

You already know that it is important to periodically save your documents. This process helps protect your work in case of catastrophic power failure or inadvertent massive edits (such as those imposed by an errant macro). When your document is small, saving to disk can be done very quickly. As your document grows, or as you start saving your document to storage devices that aren't that speedy, saving can take quite a bit longer to do.

To overcome the delay normally associated with saving a document, Word uses what is known as ***background saving.*** This term simply means that Word allows you to continue working as it writes your document to disk. The benefit is that you can keep right on working as Word does its housekeeping. You can tell when a background save is taking place because an animated disk appears on the status bar. When the disk disappears, the save is complete.

You can control whether Word utilizes background saving in the following manner:

1. Display the File tab of the ribbon and then click Options. Word displays the Word Options dialog box.
2. At the left side of the dialog box click Advanced.
3. Scroll through the options until you see the Save section.
3. If the Allow Background Saves check box is selected, Word uses the background saving feature.
4. Click on OK to dismiss the Options dialog box.

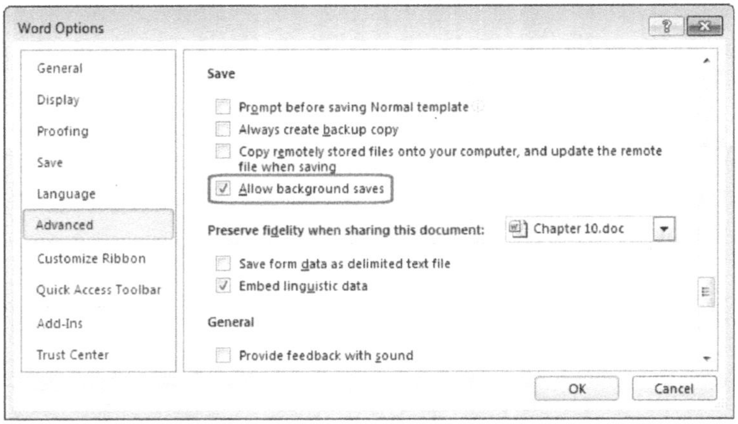

Using Filenames That Sort Properly

It is not uncommon to work on projects that require several documents. For instance, you may be working on a manual that has several parts. When you are naming files for this project, you should use names that will later sort properly when you use various functions of Word.

For instance, when you choose to open a document, you see the standard Open dialog box that shows the files in the current directory. If your files are named properly, they will always appear in order on the list. I ensure this by starting all files related to a project with two digits that represent the order in which the document appears in the project. For instance, if the project entails an introduction, three sections, and an appendix, then the files may be named as follows:

```
01 Introduction.docx
02 Section 01.docx
03 Section 02.docx
04 Section 03.docx
05 Appendix.docx
```

Without the leading two digits, of course, the files would be displayed in the dialog box in an entirely different order. For me, the above approach ensures that the files are always in the logical order I need. The only drawback is that if I need to add a new section (such as Section 04), then I need to change the number of the 05 Appendix.docx file to 06 Appendix.docx so that the new section can become 05 Section 04.docx. This renaming is a small price to pay to keep my files in the desired order.

Periodically Deleting TMP Files

As you work with Word, you may decide to turn on the AutoSave function, which causes Word to save your changes periodically. In saving these changes to disk, temporary disk files are created. (Makes sense, right?) Unfortunately, it is possible for these files to remain on the disk after you finish using Word.

Once you exit out of Word, it is OK to delete any TMP files or ASD files that may remain on your disk. Take a look in your TEMP directory and in any directories where you store Word documents. If you find any such files, feel free to delete them, thereby freeing up space on your hard drive.

You can find more information about how Word uses temporary files at this Knowledge Base page:

```
http://support.microsoft.com/?kbid=211632
```

Made in United States
Cleveland, OH
05 November 2026

47668305R00129